Concrete

RET

CTURE

JSE

CONCRETE

ITS
MANUFACTURE
AND USE

The Koehring Paver with distributing spout and road wheels.

Published by
KOEHRING COMPANY
Milwaukee, Wisconsin

The Koehring Construction Mixer with steam engine, boiler and power charging skip.

4

TABLE OF CONTENTS

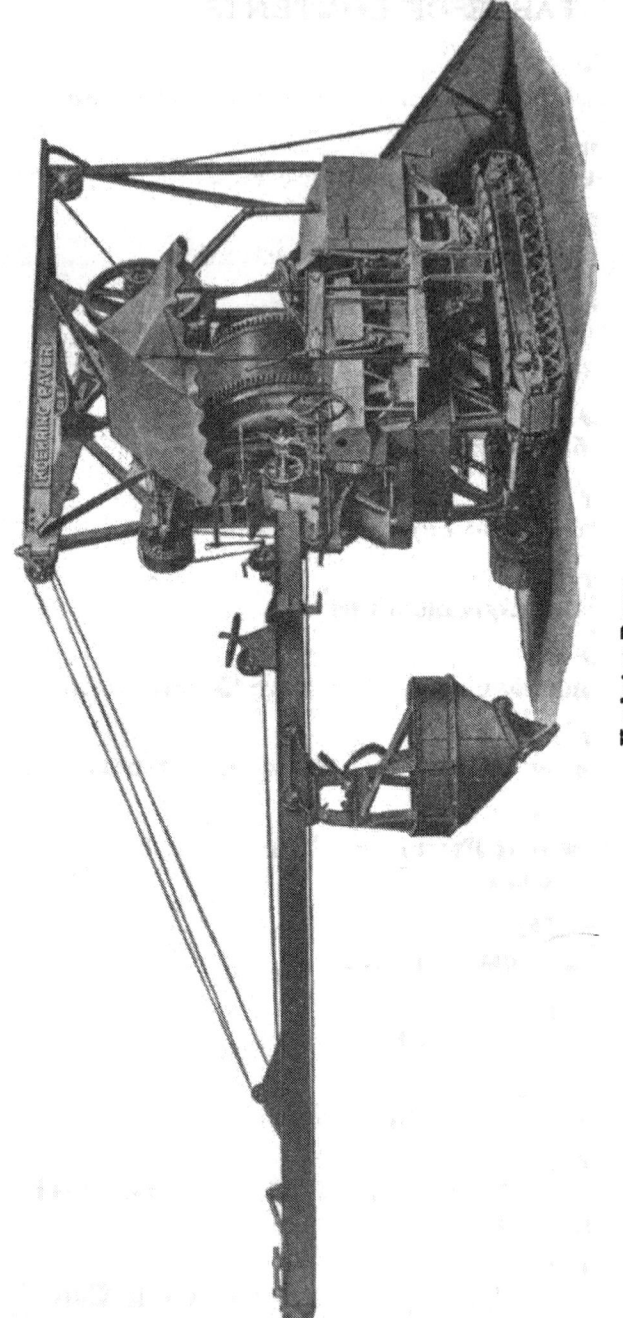

Koehring Paver

With full length multiplane traction, and distributing boom and bucket.

INTRODUCTION

The opportunities of a manufacturer of construction equipment to be of assistance to the inspector, superintendent, foreman and engineer are not numerous. Yet we have felt that there was a distinct need for a handbook that could be carried in the pocket and which would contain information concerning the best practice in the manufacture and use of concrete.

The word "manufacture" is here employed in its true sense—"To make from raw materials by any means into a form suitable for use." The construction engineer, whether he represents a contractor, or a private or government owner, who combines cement, sand, stone and water, by mixing them in a concrete mixer is as truly a manufacturer as he who combines steel, cast iron and bronze in the construction of equipment. In the same degree that the mixer manufacturer must fit all of the constituent parts to make the finished machine, so must the concrete manufacturer control the materials entering his product and the methods employed in their combination and use.

In presenting this little book no attempt is made to have it serve as a text book, nor to prescribe formulas or rules. It was compiled as a book of reference, of sound engineering practice, in concise and easily readable form. It is not a finished exposition of methods employed in all types of construction work in which concrete is used, and could not be made complete due largely to the ever changing conditions encountered.

Accept the book, therefore, in the spirit in which it has been prepared—a guide to the construction man who manufactures and places concrete in America's permanent structures, and a suggestion for the better care of equipment.

KOEHRING COMPANY

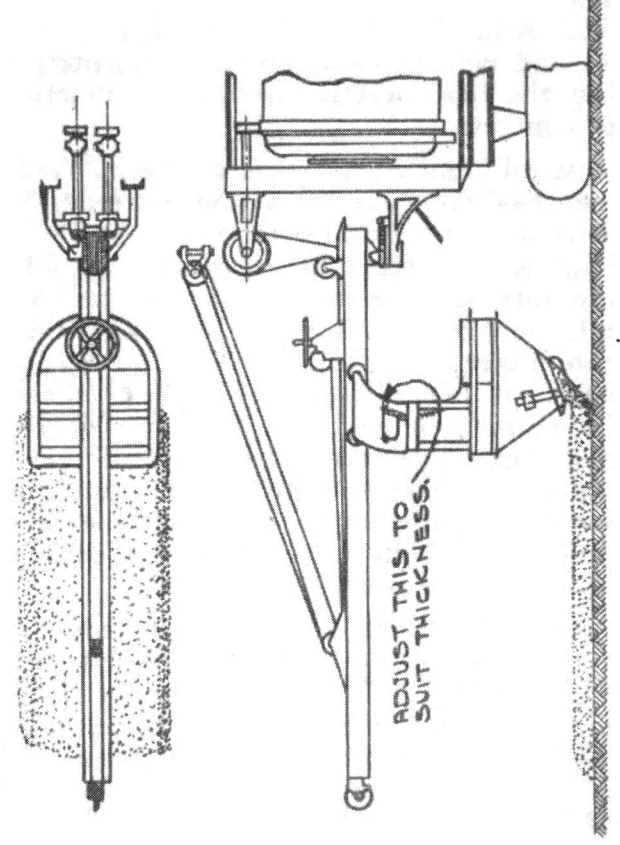

SPREAD YOUR CONCRETE THIS WAY

WITH A KOEHRING PAVER

ADJUST THIS TO SUIT THICKNESS.

8

CHAPTER 1

FIELD OPERATIONS IN CONCRETE CONSTRUCTION

Careful Planning Means Economical Completion of Project

The owner for whom the work is being done, be he an individual, a corporation, or a government agency, is interested in quality, speed and cost. The contractor and the construction superintendent are interested in cost, speed and quality. The sequence of these items is in accordance with their relative importance to the two parties to the contract.

Both are interested in each but to a different degree. The owner, whose money pays for the work, desires a structure on which depreciation will be negligible, completed in as short a time as possible after the decision has been made to go ahead, and at as low cost as is consistent with quality. Having decided to make the expenditure, his whole thought is of quality and the time at which he can put the building to use. On the other hand, the contractor having given an estimate of the cost of construction, is vitally interested in keeping the cost below that figure. To him speed means the ability to obtain his reward or profit at an earlier date, provided it can be done at equal cost with that obtained by a little slower progress. The element of interest on borrowed money must be offset against larger payrolls or a greater expenditure for plant. But even though the viewpoints of the two contracting parties are not exactly alike they are closely allied. Both are interested in the relation between cost, speed, quality and mechanical equipment.

It is not possible to give any fixed rule for field organization or for the plant required for a project. The character of the enterprise, its location, relation to railroad facilities and to other structures, and local traffic conditions over which the contractor has no control affect the decision as to the most economical method of construction. It is evident, however, that the success of the enterprise depends entirely upon the careful planning of the method by which the work is prosecuted, and the proper coordination of all items entering into it. These items will be taken up in accordance with their relation to the placing of the concrete.

Receipt of Materials and Plant for Handling

In the construction of every project, whether it be a highway, building, dam, or bridge, a considerable amount of time must be given to the preparations of the site. During this time receipt of materials for the concrete should be arranged for and delivery commenced. These materials include cement, sand, stone, reinforcing steel, and lumber and steel for forms. Finally the construction plant should come on the ground. The construction plant has been given last place, not because it is least important, but to emphasize the fact that without the materials of construction available the plant can do nothing. Many contractors have found themselves fully equipped with expensive plants but with no materials to keep them busy.

This condition can be overcome only by storing a considerable quantity of materials before the plant is erected or provided. Aggregates and cement must be on hand in such quantities as to insure continuous operation of the concrete plant. The cost of storage and rehandling is more than offset by the wages paid to skilled mechanics and superintendents if the work is delayed after the organization is in the field.

Storage of aggregates and cement may be at the point of deposit or at the point of delivery from boat or car. It is probable that the contractor can protect himself against delays due to truck breakdowns while hauling material from the point of delivery to the point of deposit, yet it is impossible with the present condition of railroad transportation to insure regular delivery from the source of supply to the railroad siding. Such storage means increased handling. Practically all materials must be rehandled at least once. Such factors as interest on money and the tying up of working capital must be taken into consideration. This entails more expense to the contractor unless he can reduce cost at some other point to justify it.

The first essential for economical handling of aggregates is the substitution of mechanical equipment for man power. Three types of equipment are available: first, bucket or skip elevators carrying material from bin under the track to the stock pile; second, locomotive cranes operating on a track, or on wheels or multiple traction, handling materials from cars into piles; and third, a derrick equipped with clam-shell bucket. The quantity of materials to be handled per day, ability to reload materials into transportation units and the quantity of storage required will determine the style of equipment best suited. In selecting this equipment the quantity of materials to be handled must be kept constantly in mind. It will be remembered that the aggregates are received in a loose state and as a result the clamshell must handle 1½ yards of aggregate for each yard of concrete. Aggregate must be handled at least twice, making necessary the use of equipment with a capacity, measured in loose materials, of at least three times that of the maximum quantity of concrete placed each day. The unloading plant is the first of those items which go to make up a perfectly balanced construction plant.

Amount of Concrete to be Placed

In laying out the remainder of the plant the one item of importance is the quantity of concrete to be placed. When this is determined it becomes a simple matter to determine the relation between plant charge and labor charge. The decision will depend upon the cost of plant, the cost of freight from the contractor's headquarters or from the point of manufacture to the point of installation, additional costs of removal, maintenance and repairs. If 25,000 cubic yards of concrete are to be placed, the allowable plant charge against the job can and should be much

11

greater than if only 500 cubic yards were to be handled. Not only should the total cost be more but the plant expenditure chargeable to the undertaking can be more per cubic yard of concrete, in order that the number of men may be reduced, thereby allowing more rapid completion of the contract even with a scarcity of men.

Character of Plant

It has been found advisable in many instances to divide the work into two distinct parts, that handled by small portable mixers and that placed by the large central mixing plant. For the preliminary work such as footings or foundation walls where the quantity of concrete required does not justify the operation of a large plant a 7-foot portable mixer, taking a one-sack batch of a 1:2:4 or a 1:3:6 mix, will prove most satisfactory. This type of machine can be set up at any convenient point, the aggregates delivered to the point of deposit and the concrete put in at lower cost than by any other method. Upon the completion of the job there will always be a lot of odds and ends that are in need of similar treatment.

Mixer Most Important Piece of Plant

The concrete mixer, because of its ability to turn out the element for which payment is received, namely, concrete, is the most important unit of the construction plant. With the increased use of concrete construction and the development of large and expensive placing plants, the ability of the concrete mixed to stand up under hard usage is a big item in making the decision for the adoption of a given machine. Delays due to breakdowns mean excessive cost. Continuity of operation spells profits. This demands heavy duty construction—a machine built of the best materials available—by competent workmen, with careful supervision.

Organization of Remainder of Plant Around Mixer

The remainder of the construction organization must be built around the mixer. Its capacity in cubic yards per day will depend largely on the equipment with which it is surrounded. This not only requires that the hoisting and placing equipment be able to take care of the output but that every link in the chain from the receipt of the cement and aggregates through the transportation, the bending of the steel, and the erection of the forms be planned entirely on the maximum mixer capacity. A little item, such as the size of the water line feeding the mixer may result in slowing down the operation from five to ten per cent. With a large plant charge constantly going on, these items must be given consideration. It must always be remembered that no pay is received except for concrete in place, and that unless the mixer is doing its work, that pay stops.

Character of Placing Equipment

The type of placing equipment and the justifiable expenditure will depend entirely on the size and character of the work. Naturally, the same type of equipment cannot be used on a highway where the concrete is placed in a thin, comparatively narrow ribbon as would be used on a building or on heavy dam construction.

For building work three general types of placing plants have been used. The first consists of a tower up which is hoisted a concrete bucket. From this tower the concrete is distributed by gravity through chutes to the point of deposit. The same type of tower is used in the second plan, together with a floor hopper from which the materials are discharged into concrete buggies or barrows. The third method consists in the use of material elevators, the concrete being elevated in carts or barrows to the point of distribution and then wheeled to place. Three items are essential regardless of the system used; first, the concrete must be placed without segregation of the particles; second, there must be control of the amount of water consistent with maintaining the flowability of concrete; and third, the concrete must be of such quality that it will flow around the reinforcing steel. Where spouts are used it is essential that the material be fed to the spouts slowly from hoppers, so that there is a constant stream of concrete. By doing this it will be found possible to use a mix which is easily handled in the forms and with which there will be no segregation.

On construction work, such as retaining walls where there is comparatively little concrete per linear foot it is sometimes desirable to set up a central mixing plant and haul the concrete for distances not to exceed 1,000 feet in each direction. If this is done, care must be taken to see that the aggregate is thoroughly wet before going into the mixer or that the concrete is mixed for a considerable length of time and that the amount of water used is kept to a minimum. Otherwise, there will be segregation causing difficulty in placing the concrete and irregular distribution of cement and aggregate in the forms.

The distribution of concrete in highway construction has been practically standardized by the adoption of the end charging paving mixer equipped with distributing boom and bucket. After being mixed the concrete is discharged from the drum into the bucket, a full batch at one time, run out on the boom and dumped on the grade. The adoption of the bottom dump bucket has made possible the depositing of concrete in place without segregation.

Quality of Concrete Desired

The character of the work to be done will determine the quality of concrete desired. It may be for a highway, for mass work or for heavily reinforced sections. Reference to table No. 1 on pages 28 and 29 shows the recommended proportions for each character of work. Peculiar grading of aggregates or the availability of certain sizes of materials may take it desirable to vary from these arbitary proportions. Before definitely determining the mix to be used on a large project, it will be found desirable to have the aggregates examined to determine their physical properties and to make compressive tests to determine the strength of concrete obtained by use of the various proportions.

Uniform Strength Demands Uniform Consistency of Concrete

The quality of concrete is dependent largely on the uniform control of the quantity of water. It is evident that the same consistency cannot be used for thin, reinforced sections as would be the case in highway construction. Yet, having adopted a consistency it is essential that this be strictly adhered to. This is easily done by measuring the water with an automatic water measuring tank with which the mixer can be equipped. The quantity of water in the aggregate will affect the amount necessary to be put into a batch, and although this water content of the aggregate will vary from day to day, depending on weather conditions, it will not be found necessary often to change the water control. It is readily appreciated that if one batch of concrete is flooded with water and the next comes out dry that the strength of the two will not be the same, as the distribution of cement will not be alike throughout the two batches.

Placing Concrete in Forms

The concrete having been brought to the point of deposit, it is essential that it be placed with care, first, in order that there may be no voids between the reinforcing steel and the concrete, thereby insuring proper bond, and second, that no honeycombed spots appear when the forms are removed. This is possible by careful spading at the sides of the form and spading or rodding the concrete sufficiently to make sure that there are no porous places left in the concrete. Satisfactory surfaces are much more readily obtained by proper spading of a plastic concrete than by trying to place the concrete when so wet that segregation results. Attention should be given to the forcing of the concrete around the reinforcing steel particularly in long, narrow or deep girders which are heavily reinforced. Reference to table 1 on pages 28 and 29 shows that in this type of construction it is desirable to use a comparatively small aggregate, so that the concrete can be placed around the reinforcing steel. Even with this aggregate, however, there is a tendency for the stones to bunch, thus forming porous spots.

Too much water, improper mixing, and improper methods of handling concrete result in laitance. Laitance consists of the finely divided dirt, silt and a certain amount of the cement, which being comparatively light and having no adhesive quality, flow to the surface. This may be generally obviated by reducing the quantity of water to the point where there is no large excess on the surface. Laitance has no strength and if allowed to accumulate will make a weak, porous layer in the structure. Special care should be taken in the concreting of wing walls for bridges and culverts when the wall is sloped to the outer end, thereby making it difficult to force the concrete into place at the end of the form. Often concrete will be allowed to fill the center of the abutment with the idea that, like water, it will seek its own level and fill up the end of the wing wall form. What happens actually is that the lighter particles, including the excess water, flow into this portion of the form and remain there until the forms are removed. If the laitance is not immediately removed and the wall repaired there will be insufficient strength in this portion of the structure to withstand frost action, to say nothing of withstanding pressure. Where laitance occurs it should be immediately cleaned from the surface before a new layer of concrete is put into place.

Curing

Reference has been made elsewhere to the necessity of proper curing of concrete, yet it may not be out of place at this point to refer to the responsibilities which this particular work puts upon the contractor. In country road construction where the concrete can be readily cured either by ponding or by wetting of an earth covering, the expense can be very readily calculated, the only additional equipment being a little larger pipe line than will otherwise be required and possibly a little larger pump. On concrete floors in buildings the question of proper curing is a much more difficult matter. Two methods have been largely employed, the first one using sawdust which is kept wet and the second method covering the floor with sand. Both of these have their drawbacks due to the expense of getting the material in place and removing it. However, the results obtained by curing more than justify the expense. It is suggested that every contractor put into his bid an item to cover curing of concrete in floors and roof slabs, as well as all pavements.

Strength

All of the foregoing items have an effect upon strength, and it is probable that in the future we will be able to increase materially our present standard for highway construction because of high strength concrete made available by more careful control of the materials entering the mixer. However, at the present time this art has not been carried far enough so that we can do no more than pay very strict attention to all these details. The item of strength is one that is of more than passing interest. True, a building may be designed for concrete having a strength of only 2,000 pounds per square inch but may inadverdently be so loaded in spots as to throw a higher stress upon it. Floors which are subject to heavy trucking, industrial driveways, and concrete pavements all must necessarily withstand great impact as well as compression. It is evident that there is a relation between strength and resistance to wear, although other factors, such as quality of aggregates materially affect this relation.

Effect of Design of Mixer on Quality of Concrete

The development of the concrete mixer has been the result of a demand on the part of contractors for a more economical way of mixing concrete, on the one hand, and a demand on the part of engineers and owners for a more uniform and stronger concrete, on the other. It is interesting to note that by far the largest part of experimental work on concrete done in laboratories has been done with concrete mixed by hand, but tests in the field show that uniformity of product is available only when properly mixed in a mixer of proper design.

Batch concrete mixers can roughly be divided into two general types. First, those that mix by a churning action only, and second, those having an action wherein the materials are alternately scattered and brought together. In the mixing of any material it is evident to obtain maximum mixing in the minimum time it is necessary to dislodge each particle of material from the particle with which it was in contact and then recombine it with another particle. In this way only can there be a true mixing action.

In the mixing of concrete there are four materials which are placed in the drum in their natural state and which must be discharged from the drum in a homogeneous, uniform mass. These materials are cement, sand, stone and water. The sand and stone are more or less of the same general character and would take their own positions with comparatively little mixing. However, it is essential that the cement and water be formed into a cement paste which thoroughly covers all of the particles of stone and sand making a matrix to hold the entire mass in place. In order that this may be done in a short time (from 1 minute to 1½ minutes) there must be a large number of actions, scattering and bringing the materials together. If, on the other hand, the materials are simply carried to the top of the drum and dropped into the mass as a whole there can be very little mixing. If a swing chute, pivoted on the inside of the drum, so as to allow it to stand at a steep angle and to throw the materials which are carried up in the pickup buckets over to the charging side, is inserted into the mixer, an entirely different mixing action takes place. The materials having been placed in the charging side of the drum, the throw-over blades, which are attached to the shell diagonally to the axis of the drum, cut through the material, moving a portion of it forward and over the top of the vane. This has a scattering effect on those particles and a mixing action as those particles strike other particles. As the drum continues to revolve the concrete is carried by these blades into the pick-up bucket, there to be carried up to the top of the drum and thrown onto the inverted swing chute. While dropping through space the particles have a tendency to separate. The direction is then quickly changed as the material strikes the swing chute, so that the concrete is spread over the width of the chute in a thin sheet prior to being returned to the charging side of the mixer to go through the remixing action again.

The water should be put into the mixer at the same time as the other materials. This is desirable in order that the mixing of the cement, water and aggregate may go on simultaneously. Incidentally, the placing of water at the same time as other materials keeps the blades of the mixer clean thereby protecting against clogging. If, on the other hand, water is put into the drum ahead of other materials the wet buckets and throw-over blades are struck by the dry materials, resulting in a tendency for the cement to stick to these parts.

It is reasonable to believe that the quality of the concrete will be materially affected by the amount of mixing that it receives. It is not surprising, therefore, that concrete of greater strength can be obtained in the same time in a machine with a complex mixing action than in a machine in which material is simply churned around. Sufficient tests are not available to make a definite statement as to the desirability of adopting a given time of mixing as standard for all machines. In order to obtain generally satisfactory results, however, engineers have adopted a one minute mix as a standard minimum.

Aggregate Control

The workability of concrete can be materially affected not only by the size, but also by the character, of aggregate. For instance, if a very porous aggregate is used it will be found difficult to handle the mixed concrete unless the aggregate is either soaked before entering the mixer or the materials are allowed to remain in the mixer for a considerable length of time. If this is not done, the water in the concrete will be quickly absorbed with the result that there will not be a sufficient amount to lubricate the mixture. A slight variation in the relation between the quantity of sand and stone may materially help the workability of certain materials. This is desirable where a coarse aggregate of one size is being used to add to a second coarse aggregate to make the mixer more workable.

Water Control

The grading and character of the aggregates, and the proportions affect the amount of water required to make a plastic, workable mix. As the quality of concrete is dependent upon a uniform quantity of water for each batch it is essential that the quantity of water be mechanically controlled.

Control of Amount of Mixing

The amount of mixing required to obtain a desired strength using the same amount of cement, aggregate and water will always be the same for a given design of machine. It may not be the same for another concrete mixer. This, as explained above, will depend entirely upon the mixing action. In order that the amount of mixing may be absolutely uniform in all instances a batchmeter has been developed which mechanically controls the number of revolutions or time of mixing for each batch. This instrument is of two types. The first is driven from the mixer drum and, knowing the speed of the drum, the time is mechanically interpolated to a dial. The second is operated by an escapement so that it measures elapsed time. The control which can be set at any point between three seconds and three minutes releases the locking mechanism allowing the discharge of the mixer and announcing the fact by the ringing of a bell.

Prior to the adoption of this mechanism on paving mixers many had not appreciated the economy of building an organization around time control. Experience has shown, however, that, knowing the time of mix, the speed with which the concrete is discharged, and the speed of charging the mixer, it is possible to develop the organization necessary to get maximum output from the machine. The whole basis of efficiency engineering is the coordinating of time with production. The batch meter is proving a pacemaker for the crew, measuring the same amount of time for each batch.

Capacity of Mixing Plant

The capacity of the mixing plant is dependent upon, first, the speed of charging the mixer, second, time of mixing, third, time for discharging, fourth, delays incident to preparation of forms and receipt of materials, fifth, organization of crew, sixth, size of batch, and seventh, ability to dispose of the concrete. The successful contractor is the one who can place the greatest quantity per machine at the least cost in a season, and not the contractor who can turn out the most concrete in an hour or even in a day.

Speed of charging is controlled largely by the method of charging the mixer. A skillful operator will hoist the charging skip and have the new batch in the drum within five seconds after the last of the concrete has been discharged from the drum, even though it may take ten or twelve seconds to raise the skip.

The time required for discharging the batch is comparatively short, so that even if the concrete is held in the drum of the mixer for a full minute it will be possible to obtain thirty to forty batches of concrete per hour. Time of mixing is not such a large factor as would at first appear because there are on every project a large number of conditions which cause delay. Investigation of construction plants show that the greatest delays are due to inability to obtain materials at the proper time, inability to get the forms in shape to receive the concrete, or inability to place the concrete after it is mixed.

Concrete mixers are built having capacities of from four to thirty-four cubic feet of concrete. Each size has its special place. It may be more economical in certain instances to use the equipment on hand for a certain time each day rather than to purchase new equipment that is exactly the right size for the job. However, as a general rule there is a tendency on the part of construction organizations to give too little thought to the size of mixer which they use on a given piece of work.

The richness of mix and the quantity of water used do not effect the capacity of the mixer plant, with the possible exception of the ability to place concrete in forms by use of towers and spouts.

Conclusion

This chapter is intended as a resume of experience based on a study of construction operations and construction plants. It has not been prepared with the idea that it is the place of a manufacturer of machinery to make definite recommendations as to the exact size of plant. There are many items which enter into such a decision and it is hoped that these have been set forth in this chapter in a way to be helpful suggestions.

CHAPTER 2

MATERIALS ENTERING CONCRETE

Concrete Aggregates

Aggregates are the inert materials, such as sand, stone screenings, pebbles, broken stone and slag, used with portland cement and water to make concrete.

Aggregates are usually classified as coarse or fine. Fine aggregate is any suitable material that will pass a No. 4 sieve or a screen having four meshes to the linear inch. Therefore, when stone screenings or other rock material is crushed so that it conforms to the above, it is regarded as sand when used in proportioning a concrete mixture.

Coarse aggregate is any suitable material such as pebbles or crushed rock of various kinds that will not pass a No. 4 sieve. Coarse aggregate may range from ¼-inch to as much as 3 inches in greatest dimension, depending upon the nature of the work for which the concrete is to be used. As a rule, however, the average maximum for most building construction is 1¼ or 1½ inches.

Concrete cannot be stronger than the materials of which it is composed. Nothing is more certain to produce unsatisfactory concrete than poor aggregates. The quality of the cement, methods of proportioning and mixing the ingredients, the amount of water used, the time of mixing and the manner of depositing concrete, all have their effect upon its density, strength and general quality, but good results cannot be expected when poor aggregates are used.

Sand, or Fine Aggregate

In the selection of sand, even greater precautions are necessary than when selecting coarse aggregates, because sand varies in physical characteristics and in general properties more widely than do the materials commonly used as coarse aggregate.

A small quantity of organic matter in sand may make it entirely unfit for use. Many impurities found in sand affect the setting of the cement, hence, the strength of the concrete. Clay in the form of a coating on the particles is injurious since it prevents the cement from contact with the surface of particles and thereby performing its bonding or binding function.

Clean Sand

If the sand is clean it should not discolor the hands. A coating of vegetable matter on sand grains appears not only to prevent the cement from hardening, but may affect it chemically. Frequently the quantity present is so small that it cannot be detected by the eye, yet may prevent the mortar in which it is used from ever reaching any appreciable strength.

A simple test for the presence of organic matter is known as the Colorimetric Test, developed by the Structural Materials Research Laboratory, Chicago. This is made as follows:

Colorimetric Test

Obtain a 12-ounce graduated prescription bottle from any drug store and fill to the 4½-ounce mark with the sand to be tested. Add to this a 3 per cent solution of caustic soda, also obtainable at any drug store, until the volume of sand and solution after shaking amounts to 7 ounces. Let this stand for twenty-four hours. At the end of this time observe the color of the liquid above the sand. If the solution is colorless or nearly so—that is, has but a pale yellowish color, the sand may be considered sufficiently free from organic impurities for any use. On the other hand, if the solution is brownish yellow in color, or darker than a pale straw color, the sand should not be used in important concrete work such as that required in roads and pavements or reinforced concrete building construction. If, in general, the color is brownish, the sand should not be used in anything but unimportant work such as footings or foundations that are not to carry heavy loads nor be subjected to frost. If the solution is dark brown in color the sand should be rejected.

This test furnishes a simple and inexpensive method of detecting the presence of such organic impurities as decayed vegetable matter. The test is being used by a large number of testing laboratories, engineers and contractors in passing on the suitability of sand for use in concrete.

Harmful Materials

The term "silt" is used to designate all foreign material which may be present in an aggregate in the form of a coating on the grains or in a finely divided state, or in the form of soft or soluble matter. Other impurities such as acids, alkalies or oils in the sand or mixing water, usually injuriously affect the quality of the concrete.

Effect of Physical Properties of Aggregate on Quality of Concrete

The hardness of aggregates grows in importance with the age of the concrete. Due to the rounded surface of the aggregate, pebble concrete one month old may be weaker than concrete made with comparatively soft broken stone, but when one year old, it may surpass in strength the broken stone concrete, because as the cement becomes harder and the bond firmer, the resistance of the aggregate to stress becomes a more important factor. The aggregate particles should offer at least as high a resistance to crushing as does the cement after attaining maximum strength. In comparing sands of the same kind, those having the highest specific gravity are likely, as a rule, to be the strongest. This applies in a general way to the comparison of different kinds of coarse aggregate also.

Grading of aggregates—that is, the relative size and quantity of the particles in the mass determines in a large degree the density of the mass. This has its effect on the strength of the concrete. However, quantity of water used, time of mixing, protection of concrete while hardening, also exert their influence, so that grading is not all-important for strength.

A sufficient quantity of fine grains is necessary in grading the material to reduce the voids, if for no other reason than to assist with the cement in increasing watertightness. An excess of fine sand has a tendency to diminish the strength of the concrete. Within reasonable limits, the strength of concrete increases with the size of the aggregates. For mass concrete, the practical maximum size is 2½ or 3 inches. In thin reinforced sections, such as floors and walls, the maximum size must be confined to particles that will enable the placing of the concrete around reinforcing steel. Generally 1 or 1¼ inches is then the preferred maximum.

The shape of aggregate particles, especially that of large aggregates, influences the strength of the mortar or concrete. Flat, elongated particles pack loosely and generally are inferior to those of more nearly cubical fracture.

Voids

Voids are air spaces between the particles and are usually referred to as a percentage of the whole. An aggregate consisting of particles all uniform in size will present the maximum of voids. This can be illustrated as follows:

Perfect spheres of equal size piled in the most compact manner, leave theoretically about 26 per cent of voids. The only requirement is that the spheres be of equal size. If, however, the spaces between the spheres in such a pile of equal size are filled with other perfect spheres of a diameter just sufficient to touch the larger spheres, the voids in the total included mass would be reduced theoretically to 20 per cent. Should this be followed up with smaller spheres, the air spaces or voids could theoretically be reduced to make the mass watertight. In practice, however, a mass of equal sized spheres will be found to contain about 44 per cent of voids.

Sampling Sand

In selecting a sample of sand for test, one should make certain that representative samples have been obtained from different parts of the pit from which material is to be used. The most representative sample is a blend of several samples taken from different parts of the pit. Whenever possible, the samples should be taken from the hoppers or from the aggregate pile after delivery upon the work.

Method of Making Void Determinations

The usual method of making void determination is by means of a graduated vessel partly filled with the sand to be tested. The amount of sand should be read and poured out. The vessel is next partly re-filled with water and the graduations read. The known amount of sand is then added to the water. The difference between the sum of the two graduations and the reading after the sand is poured into the water, divided by the amount of the sand, gives the percentage of voids.

Tests on Aggregates

Accurate tests on fine and coarse aggregate such as tensile or compressive tests can be made only in physical laboratories properly equipped for that purpose. There are certain to be apparent discrepancies between laboratory tests and field practice unless the laboratory tests are made under field conditions of proportioning and mixing. This difference is largely due to the exact control of quantity of materials entering the batch when made up in a laboratory, which is not entirely possible when the work is done in the field.

There is concrete work which has proven unsatisfactory because of the use of fine porous limestone dust or screenings. These screenings possibly were accepted after a laboratory test in which they were made up into briquettes and tested in tension, showing a strength considerably greater than the standard sand. In making the briquettes, the material was thoroughly kneaded so that there was a separation of every small particle of stone dust and a thorough coating of all grains with cement. In the field, however, this material has a tendency to ball up, with the result that the cement does not have an opportunity to surround each small particle. The concrete, therefore, never reaches full strength, is porous, and is affected by frost.

Table No. 1

TABLE OF RECOMMENDED MIXTURES
AND MAXIMUM AGGREGATE SIZES

MIXTURE AND CHARACTER OF WORK	Recommended Maximum Size of Aggregate in Inches
1:1:1 Mixture for	
The wearing course of two-course floors subject to heavy trucking, such as occurs in factories, warehouses, on loading platforms, etc.	½
1:2:3 Mixture for	
Reinforced concrete roof slabs	1
One-course concrete road, street, and alley pavements..................	3
One-course walks and barnyard pavements	1½
One-course concrete floors	1½
Fence posts..................	¾
Sills and lintels without mortar surface	¾
Watering troughs and tanks	1
Reinforced concrete columns	1
Mine Timbers	¾
Constructions subjected to water pressure, such as reservoirs, swimming pools, storage tanks, cisterns, elevator pits, vats, etc.................	1
1:2:4 Mixture for:	
Reinforced concrete walls, floors, beams, columns, and other concrete members designed in combination with steel reinforcing.................	1
Concrete for the arch ring of arch bridges and culverts.................	1½
Foundations for engines causing heavy loading, impact and vibration.................	3
Concrete work in general subject to vibration	1½
Reinforced concrete sewer pipe.................	½
1:2½:4 Mixture for	
Silo walls, grain bins, coal bins, elevators and similar structures.................	1½
Building walls above foundation, when stucco finish will not be applied.................	1½
Walls of pits or basements, exposed to moisture...	1½
Manure pits.................	1½
Dipping vats, hog wallows.................	1
Backing of concrete block.................	¾
Base of two-course road, street and alley pavements.................	3

Table No. 1

TABLE OF RECOMMENDED MIXTURES
AND MAXIMUM AGGREGATE SIZES

MIXTURE AND CHARACTER OF WORK	Recommended Maximum Size of Aggregate in inches
1:2½:5 Mixture for	
Walls above ground which are to have stucco finish	1½
Base of two-course walks, feeding floors..........	1½
Bridge abutments and wing walls, culverts, dams small retaining walls, when not reinforced......	2
Basement walls and foundations where water tightness is not essential.......................	2
Foundation for small engines..................	2
1:3:6 Mixture for	
Mass concrete—large retaining walls, heavy foundations and footings......................	3
1:1½ Mixture for	
Inside finish of water tanks, silos, and bin walls, where required, and for facing walls below ground when necessary to afford additional protection against the entrance of moisture..............	To pass through No. 8 Screen)
Back plastering of gravity retaining walls........	To pass through No. 8 Srceeen)
1:2 Mixture for	
Facing block and similar concrete products......	¼
Wearing course of two-course walks, floors subjected only to light loads, barnyard pavements, etc...................................	¼
1:2½ Mixture for	
Scratch coat of exterior plaster (cement and stucco)	To pass through No. 8 Screen)
Fence posts when coarse aggregate is not used....	¼
1:3 Mixture for	
Intermediate and finish stucco coats...........	To pass through No. 8 Screen)
Concrete block when coarse aggregate is not used.	¼
Concrete brick.............................	¼
Concrete drain tile and pipe when coarse aggregate is not used...............................	¼
Ornamental concrete products.................	¼

During the past few years a number of state highway departments have undertaken field tests with a view of ascertaining the quality of concrete obtained in the field. It has been found that a 6 by 12 cylinder is the best size specimen to use. The cylinders are made in the field, the concrete being taken from the mixer during its operation. After hardening they are shipped to the laboratory to be tested in compression and by impact.

Washing Aggregates

If the only aggregates available for use, contain some of the objectionable foreign materials already mentioned, they should be washed before using. Appreciating the importance of clean aggregates, there are many aggregate plants which now make it a practice to supply only washed aggregates for concrete construction.

Small quantities of an aggregate may be washed in trough-like devices, set at sufficient angle so that when the materials are shoveled into the upper end and a strong stream of water allowed to play down the trough, silt and clay will be removed by the tumbling, rolling, washing action as the aggregates travel toward the lower or outlet end of the trough. Where any considerable quantity of aggregates must be washed, special washing plants, usually equipped with rotary screens to size the particles are necessary.

Effect of Aggregate on Fire Resistive Qualities of Concrete

Concrete has assumed its place in building construction because of its ability to resist fire. To attain this in the greatest degree, it is necessary that the aggregates be selected for their fire resistive properties. Some aggregates that might be suited to construction where the concrete is to be exposed principally to impact or wear, would not be suitable where fire resistance is desirable. The best aggregates are siliceous sands, traprock and slag, but some grades of hard limestone have also proven particularly suitable to fire resistive construction.

Steam coal cinders also may be used, but great care should be taken to see that they are free from particles of unburned coal and contain no ash or other refuse. Steam coal cinders that are burned to a clinker are best. Both slag and trap rock have high resistance to fire and probably make the best coarse aggregate for the highest type of fire resisting construction.

Effect of Mineral Properties of Aggregate on Strength of Concrete

Sometimes the mineral properties of certain aggregates are such as to render them unfit for use in a concrete mixture. Iron ore and rocks containing some other mineral substances are not suitable: neither are slags from some smelting processes. The slag preferred is that from blast furnaces used in iron ore reduction. Where sand and gravel obtained in iron producing regions have not given satisfactory result in concrete, it has generally been due to the mineral content. The oxidation of the iron present has been known to form sulphite, so that the chemical action was powerful enough to break up the concrete through disintegration.

Acceptability of Aggregates

Cinders are used principally for concrete where very light loads are involved or to protect steel structural members. They are frequently used as aggregate for concrete in floor construction or in blocks.

Stone screenings, if from hard, durable rock, such as granite or traprock, and if well graded, clean and free from an excess of dust, may be used in place of sand with satisfactory results. As a wearing surface for floors, clean washed screenings from granite are very desirable. Owing to the difficulty of obtaining screenings that are sufficiently free from dust they should not be used unless they are first washed.

Broken concrete should never be used as aggregate. The fact that there are but few of the original sand and stone or pebble surfaces exposed, makes it almost impossible for the cement to bond broken particles of concrete together.

Chats, a product of zinc smelting, are frequently used as a concrete aggregate, particularly in the zinc mining region of Missouri where it is the principal aggregate material available. When properly combined with the necessary amount of hard sand and cement, the result is a good concrete.

Quality of Concrete not Dependent on Cement Alone

A popular supposition among many users of cement is that failures in concrete construction are caused by poor cement. Portland cement is a manufactured product, the qualities of which can be exactly controlled, and as cement manufacturers must make a product which will meet standard specification requirements, it can be realized that any of the well known brands of cement, of which there are many, must be dependable products or the manufacturers could not long stay in business.

Careful investigation of concrete failures has generally shown that aggregates, workmanship, or some factor other than the cement has been responsible for failure when such has occurred. There are, no doubt, concrete structures standing which are not what they should be because of faults of workmanship or the introduction of some bad practice.

A word of caution should be given concerning the use of so-called pit-run gravel. Almost invariably such material contains a volume of sand practically twice that of the volume of pebbles, while for good concrete, the bulk of pebbles or other coarse aggregate should be practically twice the bulk of sand. Also gravel pits frequently are not stripped of overlying soil before they are worked and this soil, which is usually humus, or rotted vegetable matter, drifts down the face of the pit and becomes mixed with the materials. In such a case they should not be used until washed. They should also be passed over suitable screens so that the sand and pebbles may be separated and remixed in proper proportions.

Careful tests will show that no two wagon-loads of gravel taken from the same pit are alike in contained volumes of sand and pebbles, nor have throughout the same grading of particles. Even where the natural run of bank material is fairly well graded, this grading will be considerably out of balance when the material is dug from the pit because the coarser particles drift down the face, so one load will consist largely of pebbles, while the next load will be largely of sand. Bank-run material, therefore, should be prepared for use by screening into two volumes—sand and pebbles. Even were it possible to make good concrete by disregarding the desirability of screening and reproportioning the materials, economy would dictate it because of the reduced quantity of cement required to produce a concrete of given strength.

Table No. 2

TABLE SHOWING CUBIC YARD WEIGHTS IN POUNDS,
EQUIVALENT WEIGHT IN TONS AND FRACTIONAL
NUMBER OF CUBIC YARDS PER TON

Weight of Aggregates in Pounds per Cubic yard	Equivalent Weight in Tons	Fractional Number of Cubic Yards per Ton
2,100	1.050	0.952
2,150	1.075	0.930
2,200	1.100	0.909
2,250	1.125	0.888
2,300	1.150	0.869
2,350	1.175	0.851
2,400	1.200	0.833
2,450	1.225	0.816
2,500	1.250	0.800
2,550	1.275	0.784
2,600	1.300	0.769
2,650	1.325	0.754
2,700	1.350	0.740
2,750	1.375	0.727
2,800	1.400	0.714
2,850	1.425	0.701
2,900	1.450	0.689
2,950	1.475	0.677
3,000	1.500	0.666
3,050	1.525	0.655
3,100	1.550	0.645
3,150	1.575	0.635
3,200	1.600	0.625
3,250	1.625	0.615
3,300	1.650	0.606

CHAPTER 3.

CONCRETE IN HIGHWAY CONSTRUCTION

In this chapter the discussion of the use of concrete in highway construction will be limited to the improvement of streets, roads, and alleys by paving.

Concrete has become an important factor in the construction of highways whether it is used as a material forming the entire pavement slab, as in the case of concrete pavements, or forms the foundations to support various types of surface such as brick, sheet asphalt and asphaltic concrete. When used for the entire pavement, it must be designed and built to give the type of surface required of a first class heavy duty highway. When concrete serves as a base or a foundation its principal function becomes one of carrying the load and distributing it over the subgrade.

Probably in no other field of use is concrete subjected to such severe abuse as when used for the construction of the all-concrete highway, whether this be road, street or alley. The concrete not only must be relied upon to furnish the desired surface for traffic but must resist the impact and abrasion resulting from the weight and volume of such traffic. It is evident that more exacting requirements must be made for concrete in the all-concrete pavement than in concrete used as the base or foundation for another type of wearing surface. This does not mean, however, that the concrete foundation work for any type of wearing surface can be mixed or placed in a slipshod manner.

One and Two Course Construction

As a rule concrete pavements are of one course construction. So-called two course construction is used in case the supply of local materials is of such quality that the required resistance to wear and surface impact cannot be met by a concrete wearing course in which the local aggregates are used. In other words, in two course concrete highway construction the top or wearing course contains aggregates especially selected for toughness and wear resistance.

Design of Pavements

The design of concrete pavements is still in the process of evolution. Many practices have come to be regarded as standard because this type of pavement has now had sufficient years of test to have proven the desirability of making certain requirements fundamental. Among these are drainage of foundation or subgrade, proper crown of the pavement surface, so that the water may be quickly removed, slope of shoulders to the gutters, suitable outlets for delivering water to culverts and natural channels, and associated details which render and maintain the foundation or subgrade in a suitable condition. Naturally soil conditions as well as climatic conditions may require that some one or more of these features be given particular attention.

Width is determined principally by traffic demands. With motor traffic now predominating on most highways it is evident that the minimum width should be established with particular reference to the safety of motor traffic when passing at prevailing speeds.

The thickness required to meet traffic conditions has not been standardized. Traffic has changed both in amount and volume with greater rapidity than has the knowledge of the requirements of highway construction. The same lack of standardization is apparent in the reinforcement of concrete pavements, although it is pretty well established that reinforcement under many conditions of subsoil is not only desirable but extremely advisable. The duty of the reinforcement is to prevent appreciable opening of cracks which may form.

Concrete Base

When concrete is used as a base for other types of surfacing, it may be plain or reinforced, but since the base is designed almost entirely from the standpoint of load carrying capacity the mixtures used

usually are different from the standard mixtures
used in the all-concrete pavement.

Curves

The need for utmost safety of traffic on modern
highways because of the predominance of motor
traffic has caused more attention to be given in the
past two or three years to certain requirements
of design and construction which a few years ago
were not recognized as necessary. Among these
are easy curves, super-elevated and widened, so that
motor vehicles can take them in safety without unnecessary reduction in speed.

Shoulders

Shoulders for a concrete road are provided to
allow additional room for passing, to give the entire roadway a finished appearance and to assist in
carrying away the water from the pavement. In
the case of narrow roads, shoulders are usually constructed and maintained to receive a portion of the
traffic. They are made of macadam, either water
bound or tar bound, or of gravel or of natural earth,
properly graded toward the side. If of ordinary
earth the usual practice is to endeavor, as soon as
possible to cover them with a growth of grass, except for that portion used for passing vehicles, thus
preventing wash and making maintenance easier
and less costly.

Curbs

In certain cases roads as well as streets are
bordered by curbs. This is particularly true of
curves on grades, and on stretches through cuts.
Usually such curbs are made integral with the
pavement and together with the surface of the
pavement form a gutter to carry surface water to a
permanently located outlet. The integral curb is
suitable for boulevard, avenue, business thorough-
fare, alley, driveway or road.

Materials of Construction

The materials entering into concrete used in
highway construction are the same as when used

for other structural purposes. They are cement, aggregates and water. Elsewhere the required physical properties of materials have been considered. These apply to concrete used in highway construction, whether for base or for the all-concrete pavement, except that in the base the aggregates are not exposed to the immediate contact of traffic. Table No. 10 on page 58 gives the cubic yards of concrete per linear foot and per mile of pavement for various widths and thicknesses. Table No. 11 shows the quantity of cement, sand and stone required per linear foot and per mile of road for different mixtures of concrete. The following table gives recommended thickness for concrete roads and streets.

Table No. 3

THICKNESS FOR LIGHT TRAFFIC ROADS AND STREETS

Width Feet	Sides Inches	Center Inches
10	6	8 (inside)
18	6	8
27	6	9
36	6	9 ½

Table No. 4

THICKNESS FOR HEAVY TRAFFIC ROADS AND STREETS

Width Feet	Sides Inches	Center Inches
20	8	10
30	8	11
40	8	12

Just as it has been the practice to increase the thickness of the all-concrete highway pavement, so has the tendency within the past year or two been to increase the thickness of concrete base used for the other types of wearing surface. Also the mixture for concrete bases has been made richer. In Illinois the standard mix for concrete foundation for other types of surface is 1:2:3½, and the thickness from four to six inches. The tendency in other parts of the country is also toward wider and thicker concrete pavements and concrete bases in keeping with the great increase of heavy motor traffic on all classes of highways.

Drainage

As already mentioned, much of the stability of any type of road surface depends upon the care given to draining the foundation or subgrade. The purpose of drainage is to secure and maintain uniformity of subgrade condition. It may be necessary to lay tile drains not only paralleling the pavement but beneath it. In flat country, for example, shoulder drains are generally necessary if the pavement is placed on clay soil. In case the construction of a highway necessitates the cutting of a clay or shale hill where sub-surface water is troublesome it is advisable to construct herringbone cross drains of broken stone or tile at least eighteen inches deep and one foot wide. These drains have their beginnings at the center of the road and extend to the gutter, at an angle of from thirty-five to forty-five degrees to the center line of the pavement. This construction is particularly applicable where the slope of the rock strata is such that the roadway cuts natural water-bearing seams.

Mechanical Equipment.

As the increasing highway programs of the state and federal governments developed, improved methods of handling materials and organizing the projects made their appearance. Those interested in the construction industry, including contractors, engineers and manufacturers of equipment, have developed new machines to meet the new requirements, or have adapted to highway work the machines used successfully in other lines.

The paving mixer has gone through an interesting development. As originally designed, the material was placed in a narrow skip barely wide enough to allow one wheelbarrow to dump into it. Concrete was distributed by spout or by horse drawn cart. The next prominent step was the development of the boom and distributing bucket, which was made automatic the next year. Then came the widening of the skip which allowed two men to discharge their wheelbarrows into it at the same time. Next came multiplane traction to take the place of road wheels on the traction end of the mixer. Finally, with the development of the use of industrial railroads, a derrick was added to pick the batch boxes from the cars. Today the paving mixer stands, an example of the highest type of construction equipment.

The capacity of the mixer to be used on a specific contract or to be adopted as standard by a contractor should be based upon the following: (1) the quantity of highway to be placed per year; (2) availability of sufficient materials to keep the equipment busy; (3) railroad facilities, car supply, etc.; (4) storage space for materials and availability of railroad sidings along line of the work; (5) type of hauling equipment to be used; (6) character of work to be undertaken.

Having determined the size of the machine to be adopted, all other equipment should be purchased with one idea, to keep the machine going to full capacity with a minimum expense. This does not mean a minimum first cost for equipment, but does mean a balanced plant that will reduce labor costs to a minimum consistent with capacity production.

Equipment is naturally divided according to its use into handling and storage, hauling, pumping, mixing and placing, and finishing. The relation of equipment to the economical completion of a project will be considered in the order named.

Handling Materials

Methods of handling materials from cars vary. The size of the project and the method of handling other parts of the job will affect the choice. The methods of unloading include, first, unloading by hand either into wagons or trucks, or into movable bodies hung against the side of the car from which

The Koehring Crane Excavator
with gasoline engine
and multiplane traction.

the material is dumped into trucks or wagons; second, mechanical unloading, using bucket elevators or skip hoists from pits below the track; and third, the use of a clam shell bucket on a derrick or a traction crane. The first and second of these methods have proved very successful where comparatively small amounts of materials were handled or where the material could be obtained in bottom dump gondola cars, but they do not give the required capacity in case the storage must be at the railroad station and not on the subgrade.

The crane fits the requirements so much better than the others that even at a larger initial investment it has been found an economical addition to a plant. The speed of its swing, its ability to travel alongside the piles, and thus to increase the storage capacity, are large factors in its favor.

Opinion is divided as to the effectiveness of two types of material storage units—the bin and the tunnel. Bins are less expensive in first cost, but the tunnel system has the advantage of reducing the amount of rehandling. The nature of the remainder of the plant used will have an effect upon the decision.

No very satisfactory mechanical equipment has been developed for handling cement from cars to storage and into the haulage units. For sacked cement manually operated two-wheel trucks have proved the most satisfactory. Some progress has been made in the use of mechanical appliances but at best the handling of the cement is expensive. Bulk cement has been found practicable when shipped either in box cars or in open top gondolas protected by tarpaulins. When shipped in box cars, handling is accomplished with a power scoop drawing the material through the door of the car into the boot of a conveyor. When received in gondolas the material is handled by locomotive crane or derrick equipped with clamshell bucket.

In case materials of construction are not stored on the sub-grade concentrated storage at one point or at several points along the line of the road must be adopted. Not only does this change the system of operation but it increases the amount of railroad track facilities required to complete a project. If the mixer is charged by wheelbarrows or with the mixer loader, it is not uncommon to see materials distributed for from one-half to two miles, ready to be placed in the mixer. This equals 1500 to 8000 cubic yards of aggregate or sufficient materials to carry on the operation from one week to one month. Cement storage may be available in barns along the line of the work. The storage space required at the railroad siding under this method of construction therefore is small.

If the materials are to be kept off the sub-grade, all this is changed. Facilities must be provided at the railroad for a large amount of material. It is probable that the economy resulting justifies the added expenditure for plant, yet the new condition must be studied closely to determine whether available facilities are sufficient to operate economically or whether it is necessary to increase the existing trackage or build sidings at another point.

In order to bring out more clearly the method of designing such a plant an assumed job will be discussed. On this project there is sixteen miles of sixteen foot road, six inches thick at the sides and eight inches thick at the center. Proportions specified are one part cement, one and one-half part sand and three parts stone. Further, because of the peculiar railroad situation it is necessary to haul materials for the whole sixteen miles from one point near the center of the section. Based on experience in highway work, it is reasonable to believe that there will be an average of twenty working days per month, this taking into consideration Sundays and rainy days but not considering any delays due to lack of materials. The experience gained in 1920 shows that there should be at least one month's supply of materials on hand at all times if this progress is to be maintained and that there will be times after the start of the shipping season for aggregate when two month's supply may have to be stored.

Table No. 5 shows the quantity of materials required per hour, day, week and month.

Table No. 6 gives the number of cars required per day and per month for the three sizes of paving mixers.

The basis on which these tables have been prepared is an average production of thirty batches per hour for eight working hours, six days per week. It is appreciated that with proper organization this can be pushed up to four hundred batches in a ten hour day without great difficulty or at the rate of forty batches per hour; yet considering the delays to which such a project is subjected it seems reasonable to adopt thirty batches per hour over an eight hour day as a basis on which to calculate the quantity of storage space required.

The railroad facilities will be controlled by the maximum number of cars to be received per day. As there is a tendency toward bunching it is evident that trackage must be provided for at least twice the daily requirement. The quantity of trackage required for each individual job will depend upon the switching service rendered by the railroad and the distance to yards of sufficient capacity to absorb the bunching of materials. With the average project located at a considerable distance from a large city or division point it is not probable that there will be made available more than two car deliveries per day, and materials will be placed in considerable quantity only once a day. One of these switches will come in in the morning and the other either later in the day or at night. This requires that space be made available for at least twelve cars if a 14E paver is used, eighteen cars if a 21E paver is used, and twenty-nine cars if a 32E paver is used. Some leeway must be given, so that the empty cars can be pushed out of the way. Assuming 40 feet per car, and that it is necessary to have a tail track capacity for at least fifty per cent of the cars, 720 feet of track is required for a 14E paver, 1080 feet of track for 21E paver and 1740 feet of track for a 32E paver. If switching facilities are better than here given this quantity of track may be cut in half.

Table No. 5
Minimum Quantity of Storage Required for Economical Operation of Highway Project
Mix 1-1½-3.

Machine		Quantity pef Hour	Quantity per day	Quantity per 6 Days	Quantity per 20 Days
14E	cement	30 bbls.	240 bbls.	1440 bbls.	4800 bbls.
	sand	6.66 cu.yd.	53.3 cu yd.	320 cu. yd.	1066 cu. yd.
	stone	13.33 cu.yd.	106.6 cu. yd.	640 cu. yd.	2132 cu. yd.
21E	cement	45 bbls.	360 bbls.	2160 bbls.	7200 bbls.
	sand	10 cu. yd.	80 cu. yd.	480 cu. yd.	1600 cu. yd.
	stone	20 cu. yd.	160 cu. yd.	960 cu. yd.	3200 cu. yd.
32E	cement	67.5 bbls.	540 bbls.	2890 bbls.	11800 bbls.
	sand	15 cu. yd.	120 cu. yd.	720 cu. yd.	2400 cu. yd.
	stone	30 cu. yd.	240 cu. yd.	1440 cu. yd.	4800 cu. yd

Table No. 6
Cars of Materials Required per Day and per Month for Three Sizes of Pavers
Mix 1-1½-3.
14-E PAVER

Cars per Day:	Cars per Month:
cement....................1 car	cement....................19 cars
sand.......................2 cars	sand.......................36 cars
stone......................4 cars	stone......................72 cars

21-E PAVER

Cars per Day:	Cars per Month:
cement.................1½ cars	cement....................30 cars
sand.....................2½ cars	sand.......................50 cars
stone......................5 cars	stone....................100 cars

32-E PAVER

Cars per Day:	Cars per Month:
cement................2-1/6 cars	cement....................43 cars
sand......................4 cars	sand.......................80 cars
stone......................8 cars	stone....................160 cars

Table No. 7
Minimum Trackage Required for Economical Operation

	14E	21E	32E
Minimum side track space required for daily car storage..................	280 ft.	400 ft.	600 ft.
Recommended track space for economical operation One switch per day—by railroad.........	720 ft.	1080 ft.	1740 ft.

As very few individual sidings exist with this capacity which are not in constant use, the contractor may be forced to build one for himself. If so, the expense of this must be taken into consideration at the time of submitting his bid and the total expense charged against the job.

From the above figures the necessity for a large storage area and sufficient trackage is evident.

Rehandling Materials

The handling of this material from cars to storage piles is easily accomplished with the crane, but the next step is of equal importance if the operation is to prove financially successful. How best to re-handle the materials into haulage units for transportation to the mixer is the problem. Three types of equipment are in general use,—first, fixed bins; second, portable bins on wheels; third, tunnels. The last of these has not been used extensively except for projects equipped with industrial railroad. The relative cost of the three methods is in accordance with the order in which they are given above. The economy of each is so dependent upon the amount of work to be completed that it is not practicable to discuss their relative efficiency at this time. When bins are adopted they should have a capacity of at least two hours run, so that no delays will result in case the crane is employed handling cars at the other end of the pile, and that short stoppages due to unforseen conditions can be bridged.

Below is given the recommended capacity for bins of each size of mixer.

Table No. 8
Minimum Capacity of Material Bins
Based on 2 hours supply for mixer

	14E	21E	32E
Sand	13 ½ cu. yd.	20 cu. yd.	30 cu. yd.
Stone	27 cu. yd.	40 cu. yd.	60 cu. yd.

Bins of the portable type can be mounted either on railroad or road wheels, so that they may move as required by the condition of the storage piles. This moving will actually be done seldom, as under normal conditions material will be handled from cars to bins without rehandling and storing. If the material is used from storage there will be sufficient of it within the reach of the crane to keep the operation going for at least one-half day without changing the location of the bins.

Drag scrapers operated by a light hoist have proved economical for cleaning up the piles when materials are deposited beyond the reach of the crane.

Haulage Units

Haulage units may naturally be divided into two distinct classes; first, trucks and second, industrial railroad. Ten years ago teams had a large place in hauling on highway construction work. This, however, has changed until their use is so limited that there seems to be no necessity for discussing them here.

Trucks again may be divided into heavy trucks —three tons and over,—and light trucks, of one and one-half ton capacity. In the past the tendency has been to use the five-ton truck for hauling materials. The only objection to this unit has been the tendency to cut up the grade and the difficulty of its use under anything but the best of road conditions. During the last two years, therefore, a number of contractors have adopted lighter trucks equipped with pneumatic tires. These are easily handled, can go over the road rapidly and do not cut up the grade as much as do the larger trucks. They have a big disadvantage, however, in the labor involved per ton of material hauled. About all that can be said concerning the size of truck to adopt is that it will depend entirely on the condition of the roads on which the hauling is to be done, the method of handling the work, and length of haul.

The use of industrial railroad is comparatively new to the highway construction industry. True, it was used some years ago but not extensively and it was only in 1919 that the present system of use was developed. The use of batch boxes, two boxes to a car, two-foot gauge equipment, and either gasoline or steam locomotives has become practically standard practice. At the loading station, the batch is prepared, the proper amount of cement, sand and stone being put into the batch box and hauled in its dry state to the paving mixer to be mixed and placed on the road.

The only criticism of this system is the fact that any delay in the transportation results in a direct delay to the mixer. In highway work the contractor receives no remuneration until the concrete has passed through the mixer and is in place on the road.

There are three other methods of handling materials: first by trucks direct from the railroad siding to the sub-grade, from which the material is rehandled into the mixer; second, the handling of it in the same manner from stock piles at distances of from five hundred feet to one mile; and third, the use of batch boxes on flat bed trucks, hauling over the completed concrete and transferring the batch to industrial cars for hauling along the concrete already placed but not yet sufficiently hardened to be used.

This system appears to combine the advantages of both types of haulage equipment. In the first place the truck has the opportunity to travel on a hard surfaced road. In the second place, the length of the industrial railroad (and the delays incident to an industrial railroad vary at least directly as its length) is reduced to not to exceed two miles and probably not to exceed one and one-half miles.

Mixer Plant

The type and size of paving mixer to adopt will depend almost entirely on the amount of road to be placed in one season, but will be affected by the method of charging the mixer.

The central mixing plant has been used on a number of highway projects. Yet the cost of hauling the concrete as opposed to hauling the dry batch, in addition to the other costs of this method seems to be such as to make this method less efficient than the use of a paving mixer on the sub-

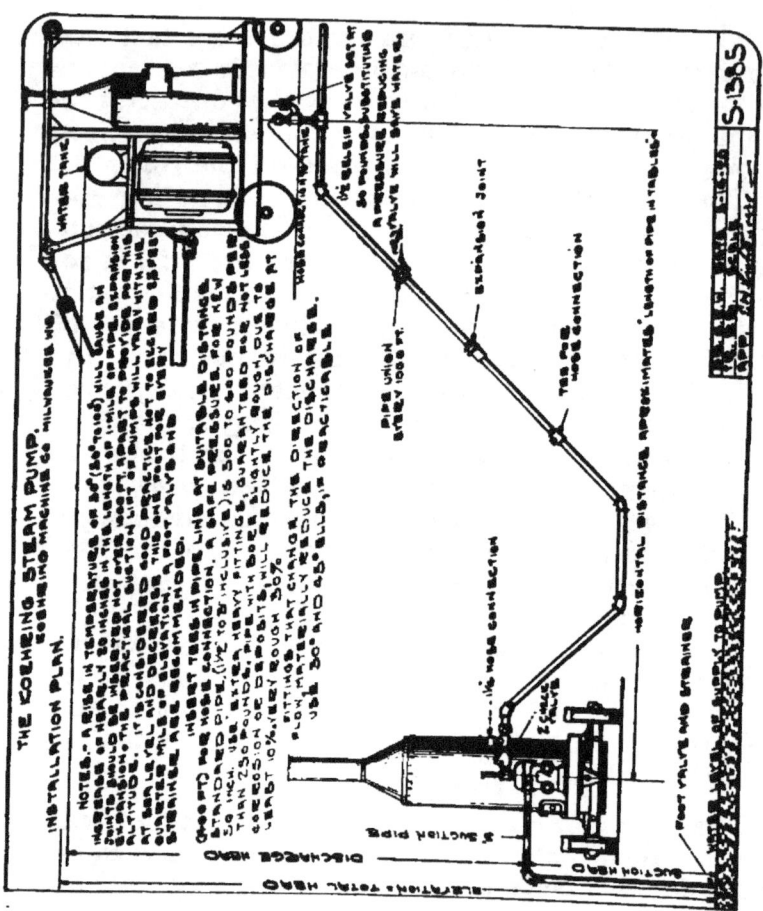

THE KOEHRING STEAM PUMP.
KOEHRING MACHINE CO., MILWAUKEE, WIS.

INSTALLATION PLAN.

grade. Undoubtedly the boom and automatic distributing bucket is the only method which has been evolved for distributing concrete in place in the proper condition to obtain the maximum strength. It is very desirable that after the concrete leaves the drum of the mixer it be put into its final place rapidly. By so doing less water can be used than would be necessary if the concrete were hauled and the resulting concrete will be denser and more homogeneous.

Of the four sizes of paving mixers it is probable that the No. 14E will remain the popular machine for any but the small or large contracts. On the small contracts the 10E will be popular while on the large contracts the support will be divided between the 21E and the 32E. The decision as to which of these sizes should be used rests entirely upon the remaining equipment and organization of the contractor.

Pump and Water Line

Rapid highway construction necessitates sufficient water at all times to insure water for the mixer, and for sprinkling and curing, as well as water for steam purposes for the other pieces of equipment on the project. Necessarily this requires a pump which automatically regulates the quantity forced into the pipe depending upon the amount used. The mistake has very often been made of using too small a pipe line on construction work. It is recommended that no pipe line less than two inches be adopted and that for distances of over one mile that two and one-half inches be adopted. The quantity of water required to carry on a project is as follows:

For mixer ....10 gallons per sq. yd.
For sprinkling and curing..25 gallons per sq. yd.
Additional for miscellaneous
    equipment ....10 gallons per sq. yd.

It is recommended that a pump of a capacity sufficient to give thirty pounds pressure at the mixer be furnished. The graph and table may be used in determining the size of pipe to use.

Table No. 9

Size of Pipe Required for Varying Length and Head

THE KOEHRING STEAM PUMP

KOEHRING MACHINE CO. MILWAUKEE, WIS.

TABLE I LENGTH OF PIPE IN FEET.

GAL'S PER MIN.	25	30	35	25	30	35	40	50	60	40	50	60	90
PIPE SIZE	1½"	1½"	1½"	2"	2"	2"	2"	2"	2"	2½"	2½"	2½"	3"

CALCULATED FOR A CONSTANT DISCHARGE PRESSURE OF 80 POUNDS PER SQ. INCH.

THE USE OF CLEAN PIPE IS ASSUMED.

ELEVATIONS HEAD IN FEET.

Use of Tables

An example of the use of Table No. 9 is as follows:

Required the size of pipe for delivery of 40 gallons of water per minute through 9800 feet of pipe to an elevation of 320 feet above the water supply. Solution—Under the heading of 40 gallons locate 9800 feet. On this line and in column marked Elevation find 200 feet. 2-inch pipes should therefore be used.

Table may also be used as follows:

Problem—How much water will the Koehring steam pump deliver through 4200 feet of 1½-inch pipe against a total head of 110 feet. To obtain this note column marked 1½-inch pipe. In this column locate 4400 and on this line in column marked Elevation read 100. Therefore the delivery under stated length will be 35 gallons per minute.

The same data can be obtained by use of Figure No. 1. Examples of the use of these figures are given as follows: Example 1—What size of pipe is required to deliver 50 gallons per minute 320 feet above and 7500 feet from the pump, suction lift being 22 feet?

Solution:—Total head equals 320 plus 22 or 342 feet. Locate 7500 on upper pipe scale and 342 on vertical scale on the right. The intersection of these two lines is near the 2½-inch pipe 50 gallon curve. Hence 2½-inch pipe is required.

EXAMPLE 2—How much will a 2-inch pipe deliver at 342 feet head and 7500 feet length?

Solution:—Locate 7500 on lower pipe scale and 342 on vertical scale on the left. The intersection of the two lines is near the 2-inch pipe 40-gallon curve. Hence nearly 40 gallons.

EXAMPLE 3—Describe the length of 2-inch pipe that can be used to deliver 40 gallons per minute against a 420-foot head.

Solution:—Note where curve marked 2 inches 40 gallons intersects with the line denoting 420-

Figure No. 1

Graph showing Capacity of Pipe for varying sizes, heads and lengths.

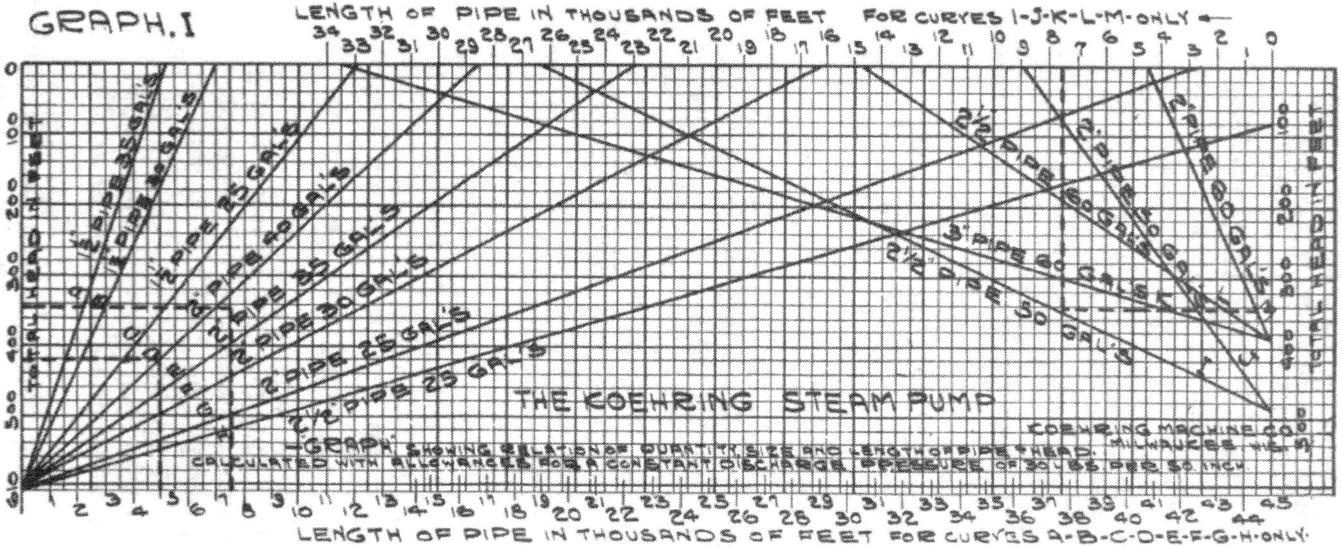

52

foot head. From that point drop down to the pipe scale and read the answer, 5000 feet.

Reference to Figure No. 1 shows the method of obtaining length of pipe, total head, and discharge head, which information is used in determining size of pipe, or the quantity of water which will flow through a pipe of a given size.

Forms and Finishing

There is little to be said concerning the type of forms to adopt. Steel forms have proven so much better than wood forms that they are recommended. With the general adoption of machine finishing it is desirable that these forms be sufficiently stiff to withstand the strains set up by the vibration as well as the weight of the machine. The finishing of concrete is one of those items which, although, it is not a major item of cost is important, as it materially affects the riding qualities of the road. In case a machine is used it is essential that the forms be kept clean, so that the wheels carrying the machine will always be at the same relative elevation and will not be forced to rise over the concrete, therefore giving a wavy surface. Where roller and belt is used care should be taken to make sure that the finishing is not completed too quickly after depositing the concrete.

Organization of Crew

The mixer capacity controls the size of the crew on all other parts of the work if the plant is properly balanced. Otherwise, the output of the mixer is controlled by the capacity of the slowest unit. The whole idea in organization of a highway construction crew is that of efficiency, time against production. Without efficiency, and efficiency means the balancing of all labor going into the production of a road, costs are going to go very high. It must be appreciated that a highway pavement is a long, thin, narrow strip for which the contractor receives a small remuneration per running foot. When it is made clear that a 14E paver has a capacity of about one and one-half to two feet of road per batch, the necessity is apparent for organizing the crew throughout in a way that each action is the most effective and can be repeated economically 3000 times in the completion of a mile of road.

The batch meter, with which all concrete mixers may be equipped, assists in this standardization of operation. This is an instrument attached to the mixer which controls the time the material remains in the drum, automatically locking the discharge chute as the charging skip reaches the top of its travel and releasing the discharge upon the expiration of the controlled time. Economy of organization can best be effected by basing the crew on the output of the mixer controlled by the meter.

Balancing of Plant

Not only is it necessary that the crew be balanced, but it must be balanced in relation to the plant, otherwise there cannot be true coordination which results in profits at the end of the project. From the receipt of materials through the handling, storage, hauling, mixing and placing, it is essential that the capacity of one machine fit the capacity of the other.

If this is done and there is proper organization of the labor, the advantage in cost is going to accrue to that contractor who has equipped himself with large, efficient machinery.

Cold Weather Work

The construction of concrete pavements is not recommended during freezing weather. It is sometimes necessary or desirable, however, to complete a small section under unfavorable weather conditions in order that an entire stretch of pavement may be thrown open to use. The fundamental precautions to be observed when concreting is done under such conditions is to maintain a certain degree of warmth in and in the presence of the concrete until it has completed early hardening and will be proof against injury from freezing.

Concrete should not be placed on a frozen sub-grade. All materials excepting the cement should be heated so that when the concrete is mixed it will have a sufficiently high temperature that the hardening process may proceed sufficiently not to be damaged by frost. It should be placed quickly and at once protected to prevent loss of the heat. Completed work can be housed in by using light frames covered with canvas, or by covering the pavement with a layer of hay or straw, a foot or more thick, after the concrete has hardened sufficiently to prevent damage to the surface from contact with this covering. A one-inch layer of sawdust with canvas over the top has been found very satisfactory when this means of protection is obtainable and the degree of cold to be protected against will not exceed three or four degrees below freezing.

Curing

One of the most important features of concrete pavement construction is proper protection of the concrete while hardening. As soon as it has been finished, the work should be covered with canvas stretched on light frames to prevent rapid evaporation of water. After hardening has advanced sufficiently to permit an earth covering being thrown on without damage to the surface, at least two inches of such covering should be applied and be kept wet by frequent sprinkling for a period of ten days or more, depending upon weather and temperature conditions. Hardening should be allowed to progress slowly and uniformly.

Where possible to arrange for it, the most satisfactory method of curing concrete pavements is to pond them. This consists of keeping them covered with about two inches of water retained by earth dams built across the pavement at suitable intervals. Flooding is generally done in the evening when the water is not needed for the mixer and is kept at a minimum depth of two inches over the crown of the pavement for at least ten days.

Maintenance of Concrete Pavements

With passage of time it becomes necessary to give attention to maintenance, although well built concrete pavements require less maintenance than any other type of pavement. Nevertheless, such maintenance as is required must be regularly and systematically given. This in general includes filling joints and cracks with tar or asphalt and repair of spots which result from local disintegration due to clay balls or other foreign materials in the concrete, or to pitting from freezing where the concrete was not properly protected while undergoing early hardening.

The materials required for maintaining concrete pavements are portland cement, sand, stone and bitumen. The last may be some one of the coal tar or asphaltic products. Demand for these for the purpose mentioned has resulted in the preparation of several standard commercial products now on the market. Care should be taken that the preparations are used in accordance with the manufacturer's recommendations, since some are intended to be applied hot and others cold. Overheating the hot preparations in general destroys their effectiveness.

But little equipment and few tools are needed for the simple maintenance work concrete pavements require. A kettle to heat tar, a can to pour it in the cracks or holes, a stiff broom for sweeping out cracks and some kind of a hooked tool, similar to a stove poker but with sharp point, may be used to scrape out compacted foreign matter from joints or cracks. For convenience in removing the tar heater, it is generally mounted on wheels, and where there is considerable mileage to maintain, the outfit is usually moved by means of a small motor truck.

Examples of Use of Tables 10 and 11

Tables 10 and 11 have been prepared with the belief that there is a need for a table giving the quantity of concrete per foot and per mile of concrete pavement and base. These tables are given for each inch of thickness, from 4 inches uniform to 8 inches at the sides and 10 inches at the center.

In table No. 11 the quantity of cement, sand and stone for each thickness, width and mixture is given. All calculations are based on quantities taken from Taylor and Thompson.

It will be noticed that in the first column three figures are given—the first is the thickness at the edge—the next the thickness at the center and the third the thickness of a uniform slab with equal cross section area.

As all pavements are built with curved crowns, the difference between the thickness given and the average of the two thicknesses is due to the greater cross sction resulting from the use of a curve. For example, a pavement 6 inches thick at the side and 8 inches thick at the center has a cross section area equal to a pavement $7\frac{1}{3}$ inches thick and not one 7 inches thick.

Example—To obtain the quantity of concrete per linear foot and per mile of a pavement 18 feet wide 7 inches thick at the edge and 8 inches thick at the center, follow along the width until 18 feet is reached. Then follow down the column until the line headed 7-8-7$\frac{2}{3}$ is reached and read .426 cubic yards per linear foot or 2249.28 cubic yards per mile. If, on the other hand, one is desirous to ascertain the quantity of cement, sand and stone required for this pavement, using a 1-2-3 mixture, refer to table No. 11—under 1-2-3 mixture—Follow down 18 foot column to point opposite 7-8-7$\frac{2}{3}$ and read .741 bbls. of cement per foot, .23 cubic yards of sand per foot and .33 cubic yards of stone per foot. In like manner the quantities of material required per mile are 3913 barrels of cement,—1192 cubic yards of sand and 1732 cubic yards of stone.

Table No. 10

CUBIC YARDS OF CONCRETE PER LINEAR FOOT AND PER MILE OF PAVEMENT

Thickness Pavement Inches				Width Pavement Feet					
Edge	Center	Average	Per Lin. Foot Per Mile	9'	10'	12'	14'	16'	18'
4	4	4	Foot	.111	.123	.148	.173	197	.222
			Mile	586.08	649.44	781.44	913.44	1040.16	1172.16
4	5	4⅔	Foot	.130	.144	.173	.201	.230	.259
			Mile	686.	760.32	913.44	1061.28	1214.40	1367.52
4	6	5⅓	Foot	.148	.164	.197	.230	263	.296
			Mile	781.44	865.92	1040.16	1214.40	1388.64	1562.88
5	5	5	Foot	.139	.154	.185	.216	.247	.278
			Mile	733.92	813.12	976.80	1140 48	1304.16	1467.84
5	6	5⅔	Foot	.157	.175	.210	.245	.280	.314
			Mile	828.96	924..	1108.80	1293.60	1478.40	1657.92
5	7	6⅓	Foot	.176	.195	.234	.274	.313	.352
			Mile	929.28	1029.60	1235.52	1446.72	1652.64	1858.56
5	8	7	Foot	.194	.216	.259	.302	.346	.389
			Mile	1024.32	1140.48	1367.52	1594.56	1826.88	2053.92
6	6	6	Foot	.167	.185	.222	.259	.296	.333
			Mile	881.76	976.80	1172.16	1367.52	1563.28	1758.24
6	7	6⅔	Foot	.185	.205	.247	288	.329	.370
			Mile	976.80	1082.40	1304.16	1520.64	1737 12	1953.60
6	8	7⅓	Foot	.204	.227	.271	.317	.362	.407
			Mile	1077.12	1198.56	1430.88	1673.76	1911.36	2148.96
6	9	8	Foot	.222	.247	.296	.346	.395	.444
			Mile	1172.16	1304.16	1562.88	1826.88	2085.60	2344.32
7	7	7	Foot	.194	.216	.259	.302	.346	.389
			Mile	1024.32	1140.48	1367.52	1594.56	1826.88	2053.92
7	8	7⅔	Foot	.213	.236	.284	.331	.378	.426
			Mile	1124.64	1246.08	1499.52	1747.68	1995.84	2249.28
7	9	8⅓	Foot	.231	.257	.309	.360	.411	.463
			Mile	1219.68	1356.96	1631.52	1900.80	2170.08	2444.64
7	10	9	Foot	.250	.278	.333	.389	.444	.500
			Mile	1320.	1467.84	1758.24	2053.92	2344.32	2640.
8	8	8	Foot	.222	.247	.296	.346	.395	.444
			Mile	1172.16	1304 16	1562.88	1826.88	2085.60	2344.32
8	9	8⅔	Foot	.241	.267	.321	.374	.428	.481
			Mile	1272.48	1409.76	1694.88	1974.72	2259.84	2539.68
8	10	9⅓	Foot	.259	.288	346	403	.461	.518
			Mile	1367.52	1520.64	1826.88	2127.84	2434.08	2735.04

Table No. 10

CUBIC YARDS OF CONCRETE PER LINEAR FOOT AND PER MILE OF PAVEMENT

Edge	Center	Average	Per Lin. Foot Per Mile	20'	22'	24'	26'	28'	30'
4	4	4	Foot	.247	.271	.296	.321	.346	.370
			Mile	1304.16	1430.88	1563.28	1694.88	1826.88	1953.60
4	5	4½	Foot	.288	.316	.345	.374	.403	.431
			Mile	1520.64	1668.48	1821.60	1974.72	2127.84	2275.68
4	6	5½	Foot	.329	.362	.395	.428	.461	.494
			Mile	1737.12	1911.36	2085.60	2259.84	2434.08	2608.32
5	5	5	Foot	.309	.339	.370	.401	.432	.463
			Mile	1631.52	1789.92	1953.60	2117.28	2280.96	2444.64
5	6	5½	Foot	.349	.384	.419	.454	.489	.524
			Mile	1842.72	2027.52	2212.32	2397.12	2581.92	2766.72
5	7	6½	Foot	.391	.430	.469	.508	.547	.586
			Mile	2064.48	2270.40	2476.32	2682.24	2888.16	3094.08
5	8	7	Foot	.432	.475	.518	.562	.605	.648
			Mile	2280.96	2508.	2735.04	2967.36	3194.40	3421.44
6	6	6	Foot	.370	.407	.444	.481	.519	.556
			Mile	1953.60	2148.96	2344.32	2539.68	2740.32	2935.68
6	7	6½	Foot	.411	.452	.493	.534	.576	.617
			Mile	2170.08	2386.56	2603.04	2819.52	3041.28	3257.76
6	8	7½	Foot	.452	.498	.543	.588	.633	.679
			Mile	2386.50	2629.44	2867.04	3104.64	3342.24	3585.12
6	9	8	Foot	.494	.543	.593	.642	.691	.741
			Mile	2608.32	2867.04	3131.04	3389.76	3648.48	3912.48
7	7	7	Foot	.432	.475	.518	.562	.605	.648
			Mile	2280.96	2508.	2735.04	2967.36	3194.40	3421.44
7	8	7½	Foot	.473	.520	.567	.615	.662	.709
			Mile	2497.44	2745.60	2993.76	3247.20	3495.36	3743.52
7	9	8¼	Foot	.514	.566	.617	.668	.720	.771
			Mile	2713.92	2988.48	3257.76	3527.04	3801.60	4070.88
7	10	9	Foot	.555	.611	.667	.722	.778	.833
			Mile	2930.40	3226.08	3521.66	3812.16	4107.84	4398.24
8	8	8	Foot	.494	.543	.593	.642	.691	.741
			Mile	2608.32	2867.04	3131.04	3389.76	3648.48	3912.48
8	9	8¼	Foot	.535	.588	.641	.695	.748	.802
			Mile	2824.80	3104.64	3384.48	3669.60	3949.44	4234.56
8	10	9¼	Foot	.575	.634	.691	.748	.806	.864
			Mile	3036.	3347.52	3648.48	3949.44	4255.68	4561.92

Table No. 11

QUANTITY OF MATERIAL REQUIRED FOR ROADS, STREETS AND ALLEYS

1—1½—3 MIX.

Quantity in Barrels of Cement—Cubic Yards of Sand—Cubic Yards of Stone.

Cement Required:—1.94 Barrels Per Cubic Yard.
Sand Required:—.42 Cubic Yards Per Cubic Yard.
Stone Required:—.84 Cubic Yards Per Cubic Yard.

Thickness in Inches				Width in Feet										
				9		10		12		14		16		18	
Edge	Center	Average		Foot	Mile	Foot	Mile	Foot	Mile	Foot	Mile	Foot	Mile	Foot	Mile
4	4	4	CEMENT	.215	1137	.238	1259	.287	1515	.335	1771	.382	2018	.431	2274
			SAND	.05	246	.05	273	.06	328	.07	383	.08	437	.09	492
			STONE	.09	492	.10	545	.12	656	.15	767	.17	874	.19	984
4	5	4½	CEMENT	.252	1331	.279	1474	.336	1771	.390	2058	.446	2356	.502	2654
			SAND	.05	288	.06	319	.07	383	.08	446	.10	510	.11	576
			STONE	.11	576	.12	638	.14	767	.17	891	.19	1020	.22	1149
4	6	5½	CEMENT	.287	1515	.318	1680	.382	2018	.446	2355	.510	2695	.574	3032
			SAND	.06	328	.07	364	.08	437	.10	510	.11	583	.12	656
			STONE	.12	656	.15	727	.17	874	.19	1020	.22	1167	.25	1313
5	5	5	CEMENT	.270	1424	.299	1577	.359	1895	.419	2213	.479	2530	.539	2848
			SAND	.06	308	.06	341	.08	410	.09	479	.10	548	.12	617
			STONE	.12	617	.13	683	.16	821	.18	958	.21	1095	.23	1233
5	6	5½	CEMENT	.305	1608	.340	1793	.407	2151	.475	2510	.543	2868	.609	3218
			SAND	.07	348	.07	388	.09	466	.10	543	.12	621	.13	696
			STONE	.13	696	.15	776	.18	932	.21	1087	.24	1242	.26	1393
5	7	6½	CEMENT	.341	1802	.378	1998	.454	2399	.532	2808	.607	3207	.683	3606
			SAND	.07	390	.08	433	.10	519	.12	608	.13	694	.15	781
			STONE	.15	780	.16	865	.20	1038	.23	1215	.26	1389	.30	1562
5	8	7	CEMENT	.378	1987	.419	2212	.502	2654	.586	3094	.671	3544	.755	3985
			SAND	.08	430	.09	479	.11	575	.13	670	.15	767	.16	863
			STONE	.16	860	.18	958	.22	1149	.25	1340	.29	1535	.33	1725
6	6	6	CEMENT	.324	1711	.359	1895	.431	2275	.502	2654	.574	3032	.646	3411
			SAND	.07	370	.08	410	.09	492	.11	575	.12	656	.14	738
			STONE	.14	741	.16	821	.19	984	.22	1149	.25	1313	.28	1477
6	7	6½	CEMENT	.359	1895	.398	2099	.479	2530	.559	2951	.638	3370	.718	3791
			SAND	.08	410	.09	454	.10	548	.12	639	.14	730	.16	821
			STONE	.16	821	.17	909	.21	1095	.24	1278	.28	1459	.31	1641

QUANTITY OF MATERIAL REQUIRED FOR ROADS, STREETS AND ALLEYS
1—1½—3 MIX.

Quantity in Barrels of Cement—Cubic Yards of Sand—Cubic Yards of Stone.

Cement Required:—1.94 Barrels Per Cubic Yard.
Sand Required:—.42 Cubic Yards Per Cubic Yard.
Stone Required:—.84 Cubic Yards Per Cubic Yard.

Thickness in Inches			Width in Feet										
			20		22		24		26		28		30	
Edge	Center	Average	Foot	Mile	Foot	Mile	Foot	Mile	Foot	Mile	Foot	Mile	Foot	Mile
4	4	4												
	CEMENT		.479	2530	.526	2776	.574	3032	.622	3288	.671	3544	.716	3791
	SAND		.10	548	.11	601	.12	656	.13	712	.15	767	.16	821
	STONE		.21	1095	.23	1202	.25	1313	.27	1424	.29	1534	.31	1641
4	5	4¾												
	CEMENT		.559	2951	.613	3236	.669	3535	.726	3832	.782	4128	.836	4415
	SAND		.12	639	.13	701	.14	765	.16	830	.17	894	.18	956
	STONE		.24	1278	.27	1401	.29	1530	.31	1659	.34	1788	.36	1912
4	6	5¼												
	CEMENT		.638	3371	.702	3707	.766	4047	.830	4384	.894	4722	.958	5060
	SAND		.14	730	.15	803	.17	876	.18	949	.19	1022	.21	1095
	STONE		.28	1459	.30	1605	.33	1752	.36	1898	.39	2045	.41	2191
5	5	5												
	CEMENT		.599	3166	.658	3473	.718	3791	.778	4108	.838	4425	.898	4743
	SAND		.13	685	.14	752	.16	821	.17	889	.18	958	.19	1027
	STONE		.26	1371	.28	1504	.31	1641	.34	1778	.36	1916	.39	2054
5	6	5¾												
	CEMENT		.677	3576	.745	3935	.813	4291	.881	4650	.949	5009	1.017	5368
	SAND		.15	774	.16	852	.18	929	.19	1007	.21	1084	.22	1162
	STONE		.29	1548	.32	1704	.35	1858	.38	2013	.41	2169	.44	2324
5	7	6¼												
	CEMENT		.759	4004	.834	4404	.910	4804	.986	5203	1.060	5603	1.137	6002
	SAND		.16	867	.18	953	.20	1040	.21	1126	.23	1213	.25	1299
	STONE		.33	1734	.36	1915	.39	2080	.43	2253	.46	2426	.49	2599
5	8	7												
	CEMENT		.838	4426	.922	4867	1.005	5306	1.090	5756	1.174	6196	1.257	6636
	SAND		.18	958	.20	1053	.22	1149	.24	1246	.25	1341	.27	1437
	STONE		.36	1916	.40	2107	.44	2297	.47	2492	.51	2683	.54	2874
6	6	6												
	CEMENT		.718	3791	.790	4169	.861	4547	.933	4928	1.007	5316	1.079	5696
	SAND		.16	821	.17	903	.19	984	.20	1067	.22	1151	.23	1233
	STONE		.31	1641	.34	1805	.37	1969	.40	2134	.44	2302	.47	2466
6	7	6¾												
	CEMENT		.797	4210	.877	4631	.956	5048	1.036	5470	1.117	5900	1.197	6321
	SAND		.17	911	.19	1003	.21	1093	.22	1184	.24	1277	.26	1368
	STONE		.35	1823	.38	2005	.41	2186	.45	2369	.48	2554	.52	2737

Table No. 11

QUANTITY OF MATERIAL REQUIRED FOR ROADS, STREETS AND ALLEYS

1—1½—3 MIX.

Quantity in Barrels of Cement—Cubic Yards of Sand—Cubic Yards of Stone.

Cement Required:—1.94 Barrels Per Cubic Yard.
Sand Required:—.42 Cubic Yard Per Cubic Yard.
Stone Required:—.84 Cubic Yards Per Cubic Yard.

Thickness in Inches			Width in Feet										
Edge	Center	Average	9		10		12		14		16		18	
			Foot	Mile	Foot	Mile	Foot	Mile	Foot	Mile	Foot	Mile	Foot	Mile
6	8	7½												
CEMENT			.396	2089	.438	2314	.525	2776	.615	3248	.702	3707	.790	4169
SAND			.09	452	.09	501	.11	601	.13	703	.15	803	.17	903
STONE			.17	905	.18	1002	.23	1202	.27	1406	.30	1605	.34	1805
6	9	8												
CEMENT			.431	2274	.480	2530	.574	3032	.671	3544	.766	4047	.861	4547
SAND			.09	492	.10	548	.12	656	.15	767	.17	876	.19	984
STONE			.19	984	.21	1095	.25	1313	.29	1535	.33	1752	.37	1969
7	7	7												
CEMENT			.388	1986	.419	2212	.502	2654	.586	3094	.671	3544	.755	3985
SAND			.08	444	.09	479	.11	575	.13	670	.15	767	.16	863
STONE			.17	887	.18	958	.22	1149	.25	1340	.29	1535	.33	1725
7	8	7½												
CEMENT			.413	2183	.458	2417	.551	2910	.642	3391	.733	3872	.826	4363
SAND			.09	473	.10	523	.12	630	.14	734	.16	838	.18	944
STONE			.18	945	.20	1047	.24	1260	.28	1468	.32	1677	.36	1889
7	9	8¼												
CEMENT			.448	2367	.499	2633	.599	3166	.698	3688	.797	4210	.898	4743
SAND			.10	512	.11	570	.13	685	.15	798	.17	911	.19	1027
STONE			.19	1025	.22	1140	.26	1371	.30	1597	.35	1823	.39	2054
7	10	9												
CEMENT			.485	2561	.539	2848	.646	3411	.755	3985	.861	4547	.970	5122
SAND			.11	554	.12	617	.14	738	.16	863	.19	984	.21	1109
STONE			.21	1109	.23	1233	.28	1477	.33	1725	.37	1969	.42	2218
8	8	8												
CEMENT			.431	2274	.480	2530	.574	3032	.671	3544	.766	4047	.861	4547
SAND			.09	492	.10	548	.12	656	.15	767	.17	876	.19	984
STONE			.19	984	.21	1095	.25	1313	.29	1535	.33	1752	.37	1969
8	9	8½												
CEMENT			.468	2468	.518	2735	.623	3288	.726	3832	.830	4384	.933	4928
SAND			.10	534	.11	592	.13	712	.16	830	.18	949	.20	1067
STONE			.20	1068	.22	1184	.27	1424	.31	1659	.36	1898	.40	2134
8	10	9½												
CEMENT			.502	2654	.559	2951	.671	3544	.782	4128	.894	4722	1.005	5306
SAND			.11	575	.12	639	.15	767	.17	894	.19	1022	.22	1149
STONE			.22	1149	.24	1278	.29	1535	.34	1788	.39	2045	.44	2297

Table No. 11

QUANTITY OF MATERIAL REQUIRED FOR ROADS, STREETS AND ALLEYS
1—1½—3 MIX.

Quantity in Barrels of Cement—Cubic Yards of Sand—Cubic Yards of Stone.

Cement Required:—1.94 Barrels Per Cubic Yard.
Sand Required:—.42 Cubic Yard Per Cubic Yard.
Stone Required:—.84 Cubic Yards Per Cubic Yard.

Edge	Center	Average	Material	20 Foot	20 Mile	22 Foot	22 Mile	24 Foot	24 Mile	26 Foot	26 Mile	28 Foot	28 Mile	30 Foot	30 Mile
6	8	7¼	CEMENT	.877	4631	.966	5100	1.053	5562	1.141	6024	1.228	6483	1.317	6955
			SAND	.19	1003	.21	1104	.23	1204	.25	1304	.27	1404	.29	1506
			STONE	.38	2005	.42	2208	.46	2408	.49	2608	.53	2807	.57	3011
6	9	8	CEMENT	.958	5060	1.053	5562	1.150	6074	1.246	6577	1.341	7077	1.438	7589
			SAND	.21	1095	.23	1204	.25	1315	.27	1424	.29	1532	.31	1643
			STONE	.41	2191	.46	2408	.50	2630	.54	2848	.58	3064	.62	3286
7	7	7	CEMENT	.838	4426	.922	4867	1.005	5306	1.090	5756	1.174	6196	1.257	6636
			SAND	.18	958	.20	1053	.22	1149	.24	1246	.25	1341	.27	1437
			STONE	.36	1916	.40	2107	.44	2297	.47	2492	.51	2683	.54	2874
7	8	7¾	CEMENT	.918	4844	1.009	5327	1.100	5808	1.193	6299	1.284	6780	1.375	7263
			SAND	.20	1049	.22	1153	.24	1257	.26	1364	.28	1468	.30	1572
			STONE	.40	2097	.44	2307	.48	2515	.52	2727	.56	2936	.60	3145
7	9	8½	CEMENT	.997	5265	1.098	5797	1.197	6321	1.296	6842	1.397	7376	1.496	7898
			SAND	.22	1140	.24	1255	.26	1368	.28	1481	.30	1597	.32	1710
			STONE	.43	2280	.48	2510	.52	2737	.56	2963	.60	3194	.65	3420
7	10	9	CEMENT	1.077	5684	1.185	6258	1.294	6833	1.401	7395	1.509	7970	1.616	8532
			SAND	.23	1231	.26	1355	.28	1479	.30	1601	.33	1725	.35	1847
			STONE	.47	2461	.51	2710	.56	2958	.61	3202	.65	3451	.70	3694
8	8	8	CEMENT	.958	5060	1.053	5562	1.150	6074	1.246	6577	1.341	7077	1.438	7589
			SAND	.21	1095	.23	1204	.25	1315	.27	1424	.29	1532	.31	1643
			STONE	.41	2191	.46	2408	.50	2630	.54	2848	.58	3064	.62	3286
8	9	8¾	CEMENT	1.038	5481	1.141	6024	1.244	6565	1.348	7120	1.451	7661	1.556	8216
			SAND	.22	1187	.25	1304	.27	1421	.29	1541	.31	1659	.34	1779
			STONE	.45	2373	.49	2608	.54	2843	.58	3083	.63	3317	.67	3557
8	10	9¼	CEMENT	1.116	5890	1.230	6495	1.341	7077	1.451	7661	1.564	8257	1.676	8850
			SAND	.24	1275	.27	1406	.29	1532	.31	1659	.34	1788	.36	1916
			STONE	.48	2550	.53	2812	.58	3064	.63	3317	.68	3575	.73	3832

Table No. 11

QUANTITY OF MATERIAL REQUIRED FOR ROADS, STREETS AND ALLEYS
1—2—3 MIX.

Quantity in Barrels of Cement—Cubic Yards of Sand—Cubic Yards of Stone:

Cement Required:—1.74 Barrels Per Cubic Yard.
Sand Required:—.53 Cubic Yards Per Cubic Yard.
Stone Required:—.77 Cubic Yards Per Cubic Yard.

Thickness in Inches				Width in Feet									
Edge	Center	Average		9		10		12		14		16		18	
				Foot	Mile	Foot	Mile	Foot	Mile	Foot	Mile	Foot	Mile	Foot	Mile
4	4	4	CEMENT	.193	1020	.214	1129	.258	1359	.301	1589	.343	1810	.386	2039
			SAND	.06	311	.07	344	.08	414	.09	484	.10	551	.12	621
			STONE	.09	451	.09	498	.11	601	.13	703	.15	801	.17	902
4	5	4¾	CEMENT	.226	1194	.251	1322	.301	1589	.350	1846	.400	2112	.451	2380
			SAND	.07	364	.08	403	.09	484	.11	562	.12	643	.14	725
			STONE	.10	528	.11	585	.13	703	.15	817	.18	935	.20	1053
4	6	5½	CEMENT	.258	1359	.285	1507	.343	1810	.400	2112	.458	2417	.515	2720
			SAND	.08	414	.09	459	.10	551	.12	643	.14	736	.16	828
			STONE	.11	601	.13	667	.15	801	.18	935	.20	1070	.23	1204
5	5	5	CEMENT	.242	1277	.268	1415	.322	1700	.376	1984	.430	2269	.484	2554
			SAND	.07	389	.08	431	.10	517	.11	604	.13	691	.15	778
			STONE	.11	565	.12	626	.14	752	.17	878	.19	1004	.21	1130
5	6	5¾	CEMENT	.273	1442	.305	1608	.365	1931	.426	2252	.487	2572	.546	2885
			SAND	.08	439	.09	490	.11	588	.13	686	.15	783	.17	879
			STONE	.12	638	.13	711	.16	854	.19	996	.22	1138	.24	1277
5	7	6½	CEMENT	.306	1616	.339	1792	.407	2152	.477	2518	.545	2876	.612	3235
			SAND	.09	492	.10	546	.12	655	.15	767	.17	876	.19	985
			STONE	.14	715	.15	793	.18	952	.21	1114	.24	1273	.27	1431
5	8	7	CEMENT	.338	1782	.376	1984	.451	2380	.525	2776	.602	3179	.677	3574
			SAND	.10	543	.11	604	.14	725	.16	845	.18	968	.21	1087
			STONE	.15	788	.17	878	.20	1053	.23	1228	.27	1407	.30	1582
6	6	6	CEMENT	.291	1535	.322	1700	.386	2039	.451	2380	.515	2720	.579	3059
			SAND	.09	467	.10	517	.12	621	.14	725	.16	828	.18	932
			STONE	.13	679	.14	752	.17	902	.20	1053	.23	1204	.26	1354
6	7	6¾	CEMENT	.322	1700	.357	1883	.430	2269	.501	2646	.572	3022	.644	3400
			SAND	.10	517	.11	573	.13	691	.15	806	.17	921	.20	1036
			TONE	.14	752	.16	833	.19	1004	.22	1171	.25	1337	.28	1505

Table No. 11

QUANTITY OF MATERIAL REQUIRED FOR ROADS, STREETS AND ALLEYS

1—2—3 MIX.

Quantity in Barrels of Cement—Cubic Yards of Sand—Cubic Yards of Stone.

Cement Required:—1.74 Barrels Per Cubic Yard.
Sand Required:—.53 Cubic Yards Per Cubic Yard.
Stone Required:—.77 Cubic Yards Per Cubic Yard.

Thickness in Inches (Edge / Center / Average)			20		22		24		26		28		30	
Edge	Center	Average	Foot	Mile	Foot	Mile	Foot	Mile	Foot	Mile	Foot	Mile	Foot	Mile
4	4	4												
CEMENT			.430	2269	.472	2490	.515	2720	.559	2949	.602	3179	.644	3400
SAND			.13	691	.14	758	.16	828	.17	898	.18	968	.20	1036
STONE			.19	1004	.21	1102	.23	1204	.25	1305	.27	1407	.28	1505
4	5	4½												
CEMENT			.501	2647	.550	2902	.600	3170	.651	3437	.701	3704	.750	3960
SAND			.15	806	.17	884	.18	966	.20	1047	.21	1128	.23	1206
STONE			.22	1171	.24	1284	.27	1403	.29	1521	.31	1638	.33	1753
4	6	5¼												
CEMENT			.572	3022	.630	3325	.687	3630	.745	3932	.802	4235	.860	4538
SAND			.17	921	.19	1013	.21	1106	.23	1198	.24	1290	.26	1382
STONE			.25	1337	.28	1471	.30	1606	.33	1740	.35	1874	.38	2008
5	5	5												
CEMENT			.538	2840	.590	3115	.644	3400	.698	3684	.752	3969	.806	4254
SAND			.16	865	.18	949	.20	1036	.21	1122	.23	1209	.25	1296
STONE			.24	1257	.26	1378	.28	1505	.31	1630	.33	1756	.36	1883
5	6	5½												
CEMENT			.607	3207	.668	3529	.726	3849	.790	4171	.851	4493	.912	4815
SAND			.18	977	.20	1075	.22	1172	.24	1270	.26	1368	.28	1467
STONE			.27	1419	.30	1562	.32	1703	.35	1846	.38	1888	.40	2131
5	7	6¼												
CEMENT			.680	3591	.748	3950	.816	4308	.884	4667	.952	5025	1.020	5384
SAND			.21	1094	.23	1203	.25	1312	.27	1421	.29	1531	.31	1640
STONE			.30	1589	.33	1748	.36	1907	.39	2065	.42	2224	.45	2382
5	8	7												
CEMENT			.752	3969	.827	4364	.901	4759	.978	5163	1.053	5556	1.128	5953
SAND			.23	1209	.25	1329	.27	1450	.30	1573	.32	1693	.34	1813
STONE			.33	1756	.37	1931	.40	2106	.43	2285	.47	2459	.50	2634
6	6	6												
CEMENT			.644	3400	.708	3739	.773	4079	.836	4420	.903	4768	.967	5109
SAND			.20	1036	.22	1139	.24	1242	.25	1346	.28	1452	.29	1556
STONE			.28	1505	.31	1655	.34	1805	.37	1956	.40	2110	.43	2261
6	7	6½												
CEMENT			.715	3776	.786	4153	.858	4529	.929	4907	1.002	5291	1.074	5669
SAND			.22	1150	.24	1265	.26	1380	.28	1495	.31	1612	.33	1727
STONE			.32	1671	.35	1838	.38	2004	.41	2171	.44	2342	.48	2509

Table No. 11

QUANTITY OF MATERIAL REQUIRED FOR ROADS, STREETS AND ALLEYS

1—2—3 MIX.

Quantity in Barrels of Cement—Cubic Yards of Sand—
Cubic Yards of Stone.

Cement Required:—1.74 Barrels Per Cubic Yard.
Sand Required:—.53 Cubic Yards Per Cubic Yard.
Stone Required:—.77 Cubic Yards Per Cubic Yard.

Thickness in Inches (Edge / Center / Average)			Width in Feet										
			9		10		12		14		16		18	
Edge	Center	Average	Foot	Mile	Foot	Mile	Foot	Mile	Foot	Mile	Foot	Mile	Foot	Mile
6	8	7½												
	CEMENT		.355	1874	.393	2076	.472	2490	.552	2913	.630	3325	.708	3739
	SAND		.11	571	.12	632	.14	758	.17	887	.19	1013	.22	1139
	STONE		.16	829	.17	919	.21	1102	.24	1289	.28	1471	.31	1655
6	9	8												
	CEMENT		.386	2039	.430	2269	.515	2720	.602	3179	.687	3630	.773	4079
	SAND		.12	621	.13	691	.16	828	.18	968	.21	1106	.24	1242
	STONE		.17	902	.19	1004	.23	1204	.27	1407	.30	1606	.34	1805
7	7	7												
	CEMENT		.338	1782	.376	1984	.451	2380	.525	2776	.602	3179	.677	3574
	SAND		.10	543	.11	604	.14	725	.16	845	.18	968	.21	1087
	STONE		.15	788	.17	878	.20	1053	.23	1228	.27	1407	.30	1582
7	8	7½												
	CEMENT		.371	1958	.411	2169	.494	2610	.576	3043	.658	3473	.741	3913
	SAND		.11	596	.13	660	.15	795	.18	926	.20	1058	.23	1192
	STONE		.16	866	.18	959	.22	1155	.25	1346	.29	1537	.33	1732
7	9	8½												
	CEMENT		.402	2126	.447	2364	.538	2839	.626	3310	.715	3778	.805	4256
	SAND		.12	646	.14	721	.16	866	.19	1009	.22	1151	.25	1296
	STONE		.18	939	.20	1045	.24	1256	.28	1464	.32	1671	.36	1882
7	10	9												
	CEMENT		.435	2299	.484	2554	.580	3058	.677	3575	.773	4080	.870	4594
	SAND		.13	700	.15	779	.18	932	.20	1090	.23	1240	.27	1399
	STONE		.19	1016	.21	1130	.26	1355	.30	1581	.34	1805	.39	2033
8	8	8												
	CEMENT		.386	2039	.430	2269	.515	2720	.602	3179	.687	3630	.773	4079
	SAND		.12	621	.13	691	.16	828	.18	968	.21	1106	.24	1242
	STONE		.17	902	.19	1004	.23	1204	.27	1407	.30	1606	.34	1805
8	9	8½												
	CEMENT		.419	2213	.465	2454	.559	2949	.651	3437	.745	3932	.837	4420
	SAND		.13	674	.14	747	.17	898	.20	1047	.23	1198	.25	1346
	STONE		.19	979	.21	1086	.25	1305	.29	1521	.33	1740	.37	1956
8	10	9½												
	CEMENT		.451	2383	.501	2648	.602	3181	.701	3705	.802	4239	.901	4760
	SAND		.13	726	.15	806	.18	970	.21	1129	.24	1295	.27	1451
	STONE		.20	1053	.22	1171	.27	1406	.31	1639	.35	1874	.40	2106

QUANTITY OF MATERIAL REQUIRED FOR ROADS, STREETS AND ALLEYS

1—2—3 MIX.

Quantity in Barrels of Cement—Cubic Yards of Sand—Cubic Yards of Stone.

Cement Required:—1.74 Barrels Per Cubic Yard.
Sand Required:—.53 Cubic Yards Per Cubic Yard.
Stone Required:—.77 Cubic Yards Per Cubic Yard.

Thickness in Inches			Width in Feet											
			20		22		24		26		28		30	
Edge	Center	Average	Foot	Mile	Foot	Mile	Foot	Mile	Foot	Mile	Foot	Mile	Foot	Mile
6	8	7½												
CEMENT			.786	4153	.867	4574	.945	4989	1.023	5403	1.101	5815	1.181	6238
SAND			.24	1265	.26	1393	.29	1520	.31	1646	.34	1771	.36	1900
STONE			.35	1838	.38	2024	.42	2208	.45	2391	.49	2573	.52	2760
6	9	8												
CEMENT			.860	4538	.945	4989	1.032	5448	1.117	5899	1.202	6348	1.289	6808
SAND			.26	1382	.29	1520	.31	1659	.34	1797	.37	1933	.39	2073
STONE			.38	2008	.42	2208	.46	2411	.49	2610	.53	2809	.57	3012
7	7	7												
CEMENT			.752	3969	.827	4364	.901	4759	.978	5163	1.053	5556	1.128	5953
SAND			.23	1209	.25	1329	.27	1450	.30	1573	.32	1693	.34	1813
STONE			.33	1756	.37	1931	.40	2106	.43	2285	.47	2459	.50	2634
7	8	7½												
CEMENT			.823	4345	.905	4778	.987	5210	1.070	5650	1.152	6081	1.234	6515
SAND			.25	1323	.28	1455	.30	1587	.33	1721	.35	1852	.38	1984
STONE			.36	1923	.40	2114	.44	2305	.47	2500	.51	2691	.55	2883
7	9	8½												
CEMENT			.895	4725	.985	5203	1.074	5673	1.162	6138	1.253	6618	1.342	7087
SAND			.27	1439	.30	1585	.33	1727	.35	1873	.38	2017	.41	2162
STONE			.40	2090	.44	2292	.48	2508	.51	2716	.55	2927	.59	3135
7	10	9												
CEMENT			.966	5098	1.063	5613	1.161	6127	1.256	6633	1.354	7148	1.449	7653
SAND			.29	1553	.32	1710	.35	1867	.38	2020	.41	2177	.44	2331
STONE			.43	2256	.47	2484	.51	2712	.56	2935	.60	3163	.64	3386
8	8	8												
CEMENT			.860	4538	.945	4989	1.032	5448	1.117	5899	1.202	6348	1.289	6808
SAND			.26	1382	.29	1520	.31	1659	.34	1797	.37	1933	.39	2073
STONE			.38	2008	.42	2208	.46	2411	.49	2610	.53	2809	.57	3012
8	9	8½												
CEMENT			.930	4916	1.023	5403	1.115	5896	1.209	6390	1.300	6873	1.395	7371
SAND			.28	1497	.31	1646	.34	1798	.37	1949	.40	2094	.43	2245
STONE			.41	2175	.45	2391	.49	2606	.54	2826	.58	3041	.62	3260
8	10	9½												
CEMENT			1.000	5284	1.103	5826	1.202	6350	1.301	6872	1.402	7410	1.503	7939
SAND			.30	1609	.34	1765	.37	1934	.40	2094	.43	2256	.46	2417
STONE			.44	2338	.49	2562	.53	2809	.58	3041	.62	3277	.67	3512

Table No. 11
QUANTITY OF MATERIAL REQUIRED FOR ROADS, STREETS AND ALLEYS
1—2—3½ MIX.

Cubic Yards of Concrete per Linear Foot and per Mile of Pavement
Quantity in Barrels of Cement—Cubic Yards of Sand—
Cubic Yards of Stone.
Cement Required:—1.61 Barrels Per Cubic Yard.
Sand Required:—.49 Cubic Yards Per Cubic Yard.
Stone Required:—.85 Cubic Yards Per Cubic Yard.

Thickness in Inches (Edge / Center / Average)			Width in Feet											
			9		10		12		14		16		18	
Edge	Center	Average	Foot	Mile	Foot	Mile	Foot	Mile	Foot	Mile	Foot	Mile	Foot	Mile
4	4	4												
CEMENT			.179	944	.198	1044	.238	1257	.279	1470	.317	1674	.357	1886
SAND			.05	287	.06	318	.07	383	.08	447	.10	510	.11	374
STONE			.09	498	.10	552	.13	664	.15	776	.17	884	.19	996
4	5	4¾												
CEMENT			.209	1104	.232	1224	.279	1470	.324	1708	.370	1955	.417	2202
SAND			.06	336	.07	372	.08	447	.10	520	.11	595	.13	670
STONE			.11	583	.12	646	.15	776	.17	902	.20	1032	.22	1163
4	6	5¼												
CEMENT			.238	1257	.264	1394	.317	1674	.370	1955	.423	2236	.477	2516
SAND			.07	383	.08	424	.10	510	.11	595	.13	681	.15	765
STONE			.13	664	.14	736	.17	884	.20	1034	.22	1181	.25	1328
5	5	5												
CEMENT			.224	1182	.248	1309	.298	1573	.348	1835	.398	2099	.448	2363
SAND			.07	360	.08	398	.09	479	.11	559	.12	639	.14	719
STONE			.12	624	.13	691	.16	830	.18	969	.21	1108	.24	1249
5	6	5¾												
CEMENT			.253	1335	.282	1488	.338	1785	.394	2083	.451	2380	.506	2669
SAND			.08	406	.09	453	.10	543	.12	634	.14	724	.15	812
STONE			.13	705	.15	785	.18	943	.21	1100	.24	1256	.27	1409
5	7	6¼												
CEMENT			.283	1496	.314	1658	.377	1990	.441	2330	.503	2661	.567	2993
SAND			.09	455	.10	505	.11	606	.13	709	.15	810	.17	911
STONE			.15	790	.17	876	.20	1051	.23	1230	.27	1405	.30	1580
5	8	7												
CEMENT			.312	1649	.348	1837	.417	2202	.486	2568	.557	2941	.626	3307
SAND			.10	502	.11	559	.13	670	.15	782	.17	895	.19	1006
STONE			.16	870	.18	969	.22	1163	.26	1356	.29	1553	.33	1746
6	6	6												
CEMENT			.269	1420	.298	1573	.357	1887	.417	2202	.477	2516	.536	2830
SAND			.08	432	.09	479	.11	574	.13	670	.15	766	.16	861
STONE			.14	750	.16	830	.19	996	.22	1163	.25	1329	.28	1494
6	7	6½												
CEMENT			.298	1573	.330	1742	.396	2099	.464	2449	.530	2797	.596	3146
SAND			.09	479	.10	530	.12	639	.14	745	.16	851	.18	958
STONE			.16	830	.17	920	.21	1108	.24	1293	.28	1476	.31	1662

Table No. 11

QUANTITY OF MATERIAL REQUIRED FOR ROADS, STREETS AND ALLEYS

1—2—3½ MIX.

Quantity in Barrels of Cement—Cubic Yards of Sand—Cubic Yards of Stone.

Cement Required:—1.61 Barrels Per Cubic Yard.
Sand Required:—.49 Cubic Yards Per Cubic Yard.
Stone Required:—.85 Cubic Yards Per Cubic Yard.

Thickness in Inches (Edge / Center / Average)	Width in Feet											
	20		22		24		26		28		30	
	Foot	Mile	Foot	Mile	Foot	Mile	Foot	Mile	Foot	Mile	Foot	Mile
4 4 4												
CEMENT	.398	2099	.436	2304	.476	2516	.517	2729	.557	2941	.596	3146
SAND	.12	639	.13	701	.15	766	.16	831	.17	895	.18	957
STONE	.21	1108	.23	1216	.25	1329	.27	1441	.29	1553	.31	1661
4 5 4¾												
CEMENT	.464	2449	.509	2685	.555	2933	.602	3180	.649	3426	.694	3664
SAND	.14	745	.15	817	.17	893	.18	968	.20	1043	.21	1115
STONE	.24	1293	.27	1418	.29	1549	.32	1679	.34	1809	.37	1935
4 6 5½												
CEMENT	.530	2797	.583	3077	.636	3358	.689	3639	.742	3919	.795	4199
SAND	.16	851	.18	936	.19	1022	.21	1107	.23	1193	.24	1278
STONE	.28	1476	.31	1624	.34	1773	.36	1921	.39	2069	.42	2217
5 5 5												
CEMENT	.498	2628	.546	2882	.596	3146	.646	3408	.696	3672	.745	3936
SAND	.15	800	.17	877	.18	957	.20	1037	.21	1118	.23	1198
STONE	.26	1387	.29	1522	.31	1661	.34	1799	.37	1939	.39	2078
5 6 5½												
CEMENT	.562	2967	.618	3265	:675	3561	.731	3859	.787	4157	.844	4455
SAND	.17	903	.19	994	.21	1084	.22	1175	.24	1265	.26	1356
STONE	.30	1567	.33	1724	.36	1880	.39	2037	.42	2195	.45	2352
5 7 6½												
CEMENT	.630	3323	.692	3655	.755	3986	.818	4318	.881	4650	.943	4981
SAND	.19	1011	.21	1112	.23	1213	.25	1314	.27	1415	.29	1516
STONE	.33	1754	.37	1930	.40	2105	.43	2280	.46	2455	.50	2630
5 8 7												
CEMENT	.695	3672	.765	4038	.834	4403	.905	4777	.975	5142	1.043	5508
SAND	.21	1118	.23	1229	.25	1340	.28	1454	.30	1565	.32	1676
STONE	.37	1939	.40	2132	.44	2325	.48	2522	.51	2715	.55	2908
6 6 6												
CEMENT	.596	3146	.655	3460	.715	3774	.774	4089	.836	4411	.895	4727
SAND	.18	958	.20	1053	.22	1149	.24	1245	.25	1343	.27	1439
STONE	.31	1662	.35	1827	.38	1992	.41	2159	.44	2329	.47	2496
6 7 6¾												
CEMENT	.662	3494	.728	3843	.794	4191	.860	4540	.927	4896	.993	5245
SAND	.20	1063	.22	1170	.24	1275	.26	1382	.28	1490	.30	1596
STONE	.35	1845	.38	2029	.41	2213	.45	2397	.49	2585	.52	2769

Table No. 11

QUANTITY OF MATERIAL REQUIRED FOR ROADS, STREETS AND ALLEYS

1—2—3½ MIX.

Quantity in Barrels of Cement—Cubic Yards of Sand—Cubic Yards of Stone.

Cement Required:—1.61 Barrels Per Cubic Yard.
Sand Required:—.49 Cubic Yards Per Cubic Yard.
Stone Required:—.85 Cubic Yards Per Cubic Yard.

Thickness in Inches			Width in Feet										
			9		10		12		14		16		18	
Edge	Center	Average	Foot	Mile	Foot	Mile	Foot	Mile	Foot	Mile	Foot	Mile	Foot	Mile
6	8	7½												
CEMENT			.371	1734	.365	1930	.436	2304	.510	2695	.583	3077	.655	3460
SAND			.10	527	.11	585	.13	701	.15	820	.18	936	.20	1053
STONE			.17	915	.19	1019	.23	1216	.27	1423	.31	1624	.35	1827
6	9	8												
CEMENT			.357	1887	.398	2099	.477	2516	.557	2941	.636	3358	.715	3774
SAND			.11	574	.12	639	.15	766	.17	895	.19	1022	.22	1149
STONE			.19	996	.21	1108	.25	1329	.29	1553	.34	1773	.38	1992
7	7	7												
CEMENT			.312	1649	.348	1837	.417	2202	.486	2568	.557	2941	.626	3307
SAND			.10	502	.11	559	.13	670	.15	782	.17	895	.19	1006
STONE			.16	870	.18	970	.22	1163	.26	1356	.29	1553	.33	1746
7	8	7½												
CEMENT			.343	1811	.380	2006	.457	2415	.533	2814	.609	3214	.686	3621
SAND			.10	551	.12	611	.14	735	.16	857	.19	978	.21	1102
STONE			.18	956	.20	1059	.24	1275	.28	1486	.32	1697	.36	1912
7	9	8½												
CEMENT			.372	1964	.414	2185	.497	2628	.580	3061	.662	3494	.745	3936
SAND			.11	598	.13	665	.15	800	.18	931	.20	1063	.23	1198
STONE			.20	1037	.22-	1153	.26	1387	.31	1616	.35	1845	.39	2078
7	10	9												
CEMENT			.403	2125	.447	2380	.536	2830	.626	3307	.715	3774	.805	4250
SAND			.12	647	.14	719	.16	861	.19	1006	.22	1149	.25	1294
STONE			.21	1122	.24	1248	.28	1494	.33	1746	.38	1992	.43	2244
8	8	8												
CEMENT			.357	1887	.398	2099	.477	2516	.557	2941	.636	3358	.715	3774
SAND			.11	574	.12	639	.15	766	.17	895	.19	1022	.22	1149
STONE			.19	996	.21	1108	.25	1329	.29	1553	.34	1773	.38	1992
8	9	8½												
CEMENT			.388	2048	.430	2270	.517	2729	.602	3180	.689	3639	.774	4089
SAND			.12	623	.13	691	.16	830	.18	968	.21	1107	.24	1245
STONE			.20	1081	.23	1199	.27	1441	.32	1679	.36	1921	.41	2159
8	10	9½												
CEMENT			.417	2202	.464	2449	.557	2941	.649	3426	.742	3919	.834	4403
SAND			.13	670	.14	745	.17	895	.20	1043	.23	1193	.25	1340
STONE			.22	1163	.24	1293	.29	1553	.34	1809	.39	2069	.44	2325

Table No. 11

QUANTITY OF MATERIAL REQUIRED FOR ROADS, STREETS AND ALLEYS

1—2—3½ MIX.

Quantity in Barrels of Cement—Cubic Yards of Sand—Cubic Yards of Stone.

Cement Required:—1.61 Barrels Per Cubic Yard.
Sand Required:—.49 Cubic Yards, Per Cubic Yard.
Stone Required:—.85 Cubic Yards Per Cubic Yard.

Thickness in Inches			Width in Feet											
			20		22		24		26		28		30	
Edge	Center	Average	Foot	Mile	Foot	Mile	Foot	Mile	Foot	Mile	Foot	Mile	Foot	Mile
6	8	7½												
	CEMENT		.728	3843	.802	4233	.874	4616	.947	4997	1.019	5381	1.093	5772
	SAND		.22	1170	.24	1288	.27	1405	.29	1521	.31	1638	.33	1757
	STONE		.38	2029	.42	2235	.46	2437	.50	2638	.54	2841	.58	3047
6	9	8												
	CEMENT		.795	4199	.874	4616	.955	5041	1.034	5458	1.113	5873	1.193	6298
	SAND		.24	1278	.27	1405	.29	1534	.31	1661	.34	1788	.36	1917
	STONE		.42	2217	.46	2437	.50	2661	.55	2882	.59	3101	.63	3325
7	7	7												
	CEMENT		.695	3672	.765	4038	.834	4403	.905	4777	.975	5142	1.043	5508
	SAND		.21	1118	.23	1229	.25	1340	.28	1454	.30	1565	.32	1676
	STONE		.37	1939	.40	2132	.44	2325	.48	2522	.51	2715	.55	2908
7	8	7⅜												
	CEMENT		.762	4020	.837	4421	.913	4820	.990	5228	1.066	5627	1.141	6028
	SAND		.23	1224	.25	1346	.28	1467	.30	1591	.32	1713	.35	1835
	STONE		.40	2122	.44	2334	.48	2545	.52	2760	.56	2971	.60	3182
7	9	8⅓												
	CEMENT		.828	4370	.911	4811	.993	5245	1.075	5678	1.159	6121	1.241	6554
	SAND		.25	1330	.28	1464	.30	1596	.33	1728	.35	1862	.38	1995
	STONE		.44	2307	.48	2540	.52	2769	.58	2998	.61	3232	.66	3460
7	10	9												
	CEMENT		.838	4717	.984	5194	1.074	5670	1.162	6137	1.253	6614	1.341	7081
	SAND		.27	1436	.30	1581	.33	1726	.35	1868	.38	2013	.41	2155
	STONE		.47	2491	.52	2742	57	2994	.61	3240	.66	3491	.71	3738
8	8	8												
	CEMENT		.795	4199	.874	4616	.955	5041	1.034	5458	1.113	5873	1.193	6298
	SAND		.24	1278	.27	1405	.29	1534	.31	1661	.34	1788	.36	1917
	STONE		.42	2217	.46	2437	.50	2661	.55	2882	.59	3101	.63	3325
8	9	8⅓												
	CEMENT		.861	4548	.947	4999	1.032	5448	1.119	5909	1.204	6358	1.291	6818
	SAND		.26	1384	.29	1521	.31	1658	.34	1798	.37	1935	.39	2075
	STONE		.45	2401	.50	2639	.54	2876	.59	3120	.64	3357	.68	3600
8	10	9⅓												
	CEMENT		.925	4988	1.021	5390	1.113	5873	1.204	6358	1.298	6852	1.391	7344
	SAND		.28	1488	.31	1641	.34	1788	.37	1935	.39	2081	.42	2235
	STONE		.49	2581	.54	2846	.59	3101	.64	3357	.69	3618	.73	3878

Table No. 11
QUANTITY OF MATERIAL REQUIRED FOR ROADS, STREETS AND ALLEYS
1—2—4 MIX.

Quantity in Barrels of Cement—Cubic Yards of Sand—Cubic Yards of Stone.

Cement Required:—1.48 Barrels Per Cubic Yard.
Sand Required:—.45 Cubic Yards Per Cubic Yard.
Stone Required:—.90 Cubic Yards Per Cubic Yard.

Thickness in Inches			Width in Feet										
			9		10		12		14		16		18	
Edge	Center	Average	Foot	Mile	Foot	Mile	Foot	Mile	Foot	Mile	Foot	Mile	Foot	Mile
4	4	4												
CEMENT			.164	867	.182	961	.219	1156	.256	1351	.292	1539	.329	1735
SAND			.05	264	.06	292	.07	351	.08	411	.09	468	.10	527
STONE			.10	527	.11	584	.13	703	.16	822	.18	936	.20	1055
4	5	4¾												
CEMENT			.192	1015	.213	1125	.256	1351	.297	1570	.340	1797	.383	2025
SAND			.06	309	.06	342	.08	411	.09	477	.10	546	.12	616
STONE			.12	617	.13	684	.16	822	.18	955	.21	1093	.23	1231
4	6	5½												
CEMENT			.219	1156	.243	1282	.292	1539	.340	1797	.389	2071	.438	2313
SAND			.07	351	.07	390	.09	468	.10	546	.12	625	.13	703
STONE			.13	703	.15	780	.18	936	.21	1093	.24	1250	.27	1407
5	5	5												
CEMENT			.206	1086	.227	1203	.274	1446	.320	1687	.366	1930	.411	2173
SAND			.06	330	.07	366	.08	440	.10	513	.11	587	.13	661
STONE			.13	661	.14	732	.17	879	.19	1026	.22	1174	.25	1321
5	6	5¾												
CEMENT			.232	1227	.259	1368	.311	1641	.363	1915	.414	2187	.465	2454
SAND			.07	373	.08	416	.09	499	.11	582	.13	665	.14	746
STONE			.14	746	.16	832	.19	998	.22	1165	.25	1330	.28	1492
5	7	6½												
CEMENT			.260	1375	.289	1524	.346	1829	.406	2142	.463	2446	.521	2751
SAND			.08	418	.09	464	.11	556	.12	651	.14	744	.16	837
STONE			.16	836	.18	927	.21	1112	.25	1302	.28	1488	.32	1673
5	8	7												
CEMENT			.287	1516	.320	1687	.383	2025	.447	2361	.512	2704	.576	3040
SAND			.09	461	.10	513	.12	616	.14	718	.16	822	.18	924
STONE			.17	922	.19	1026	.23	1231	.27	1436	.31	1644	.35	1849
6	6	6												
CEMENT			.247	1305	.274	1446	.329	1735	.383	2025	.438	2313	.493	2602
SAND			.08	397	.08	440	.10	527	.12	616	.13	703	.15	791
STONE			.15	794	.17	879	.20	1055	.23	1231	.27	1407	.30	1582
6	7	6¾												
CEMENT			.274	1446	.303	1601	.366	1930	.426	2251	.487	2571	.548	2892
SAND			.08	440	.09	487	.11	587	.13	684	.15	782	.17	879
STONE			.17	879	.18	974	.22	1174	.26	1369	.30	1563	.33	1759

Table No. 11
QUANTITY OF MATERIAL REQUIRED FOR ROADS, STREETS AND ALLEYS
1—2—4 MIX.

Quantity in Barrels of Cement—Cubic Yards of Sand—Cubic Yards of Stone.

Cement Required:—1.48 Barrels Per Cubic Yard.
Sand Required:—.45 Cubic Yards Per Cubic Yard.
Stone Required:—.90 Cubic Yards Per Cubic Yard.

Thickness in Inches (Edge / Center / Average)		Width in Feet										
		20		22		24		26		28		30	
		Foot	Mile	Foot	Mile	Foot	Mile	Foot	Mile	Foot	Mile	Foot	Mile
4 4 4	CEMENT	.366	1930	.401	2118	438	2313	.475	2509	.512	2704	.548	2892
	SAND	.11	587	.12	644	.13	703	.14	763	.16	822	.17	879
	STONE	.22	1174	.24	1288	.27	1407	.29	1526	.31	1644	.33	1759
4 5 4½	CEMENT	.426	2251	.468	2469	.511	2697	.554	2923	.596	3149	.638	3368
	SAND	.13	684	.14	751	.16	820	.17	889	.18	958	.19	1024
	STONE	.26	1369	.28	1501	.31	1640	.34	1778	.36	1915	.39	2048
4 6 5½	CEMENT	.487	2571	.536	2828	.585	3087	.633	3345	.682	3602	.731	3860
	SAND	.15	782	.16	860	.18	939	.19	1017	.21	1095	.22	1174
	STONE	.30	1563	.33	1720	.36	1877	.39	2034	.41	2191	.44	2347
5 5 5	CEMENT	.457	2415	.502	2649	.548	2892	.593	3133	.639	3376	.685	3619
	SAND	.14	734	15	806	.17	879	.18	953	.19	1026	.21	1100
	STONE	.28	1469	.31	1611	.33	1759	.36	1905	.39	2053	.42	2200
5 6 5½	CEMENT	.517	2728	.568	3001	.620	3274	.672	3548	.724	3821	.776	4095
	SAND	.16	829	.17	913	.19	995	.20	1079	.22	1162	.24	1245
	STONE	.31	1659	.35	1825	.38	1991	.41	2157	.44	2324	.47	2490
5 7 6½	CEMENT	.579	3055	.636	3360	.694	3664	.752	3969	.810	4274	.867	4579
	SAND	.18	929	.19	1022	.21	1114	.23	1207	.25	1300	.26	1392
	STONE	35	1858	.39	2043	.42	2228	.46	2414	.49	2599	.53	2785
5 8 7	CEMENT	.639	3376	.703	3712	.767	4048	.832	4391	.895	4727	.959	5063
	SAND	19	1026	.21	1129	.23	1231	.25	1335	.27	1437	.29	1539
	STONE	.39	2053	.43	2257	.47	2462	.51	2670	.54	2875	.58	3079
6 6 6	CEMENT	.548	2892	.602	3181	.657	3469	.712	3759	.768	4055	.823	4345
	SAND	.17	879	.18	967	.20	1055	.22	1143	.23	1233	.25	1321
	STONE	.33	1759	.37	1934	.40	2110	.43	2286	.47	2466	.50	2642
6 7 6½	CEMENT	608	3212	.669	3533	.730	3852	.790	4174	.852	4501	.913	4822
	SAND	.18	977	.20	1074	.22	1171	.24	1269	.26	1368	.28	1466
	STONE	.37	1953	.41	2148	.44	2343	.48	2538	.52	2737	.55	2932

Table No. 11
QUANTITY OF MATERIAL REQUIRED FOR ROADS, STREETS AND ALLEYS
1—2—4 MIX.

Quantity in Barrels of Cement—Cubic Yards of Sand—Cubic Yards of Stone.

Cement Required:—1.48 Barrels Per Cubic Yard.
Sand Required:—.45 Cubic Yards Per Cubic Yard.
Stone Required:—.90 Cubic Yards Per Cubic Yard.

Thickness in Inches (Edge / Center / Average)		Width in Feet											
		9		10		12		14		16		18	
		Foot	Mile	Foot	Mile	Foot	Mile	Foot	Mile	Foot	Mile	Foot	Mile
6 8 7½	CEMENT	.302	1594	.332	1766	.401	2118	.469	2478	.536	2828	.602	3181
	SAND	.09	485	.10	537	.12	644	.14	753	.16	860	.18	967
	STONE	.18	969	.20	1074	.24	1288	.29	1507	.33	1720	.37	1934
6 9 8	CEMENT	.329	1735	.366	1930	.438	2313	.512	2704	.585	3087	.657	3469
	SAND	.10	527	.11	587	.13	703	.16	822	.18	939	.20	1055
	STONE	.20	1055	.22	1174	.27	1407	.31	1643	.36	1877	.40	2110
7 7 7	CEMENT	.287	1516	.320	1687	.383	2025	.447	2361	.512	2704	.576	3040
	SAND	.09	461	.10	513	.12	616	.14	718	.16	822	.18	924
	STONE	.17	922	.19	1026	.23	1231	.27	1436	.31	1644	.35	1849
7 8 7½	CEMENT	.315	1665	.349	1844	.420	2220	.490	2587	.559	2954	.630	3329
	SAND	.10	506	.11	561	.13	675	.15	787	.17	898	.19	1012
	STONE	.19	1012	.21	1121	.26	1350	.30	1573	.34	1796	.38	2024
7 9 8½	CEMENT	.342	1806	.381	2008	.457	2415	.533	2813	.608	3212	.685	3619
	SAND	.10	549	.12	611	.14	734	.16	855	.18	977	.21	1100
	STONE	.21	1098	.23	1221	.28	1469	.32	1711	.37	1953	.42	2201
7 10 9	CEMENT	.370	1954	.411	2173	.493	2602	.576	3040	.657	3469	.740	3907
	SAND	.11	594	.13	661	.15	791	.18	924	.20	1055	.23	1188
	STONE	.23	1188	.25	1321	.30	1582	.35	1849	.40	2110	.45	2376
8 8 8	CEMENT	.329	1735	.366	1930	.438	2313	.512	2704	.585	3087	.657	3469
	SAND	.10	527	.11	587	.13	703	.16	822	.18	939	.20	1055
	STONE	.20	1055	.22	1174	.27	1407	.31	1643	.36	1877	.40	2110
8 9 8½	CEMENT	.357	1883	.395	2087	.475	2509	.554	2923	.633	3353	.712	3759
	SAND	.11	572	.12	635	.14	763	.17	889	.19	1017	.22	1143
	STONE	.22	1145	.24	1269	.29	1526	.34	1778	.39	2034	.43	2286
8 10 9½	CEMENT	.383	2025	.426	2251	.512	2704	.596	3149	.682	3602	.767	4048
	SAND	.12	616	.13	684	.16	822	.18	958	.21	1095	.23	1231
	STONE	.23	1231	.26	1369	.31	1644	.36	1915	.41	2191	.47	2462

Table No. 11
QUANTITY OF MATERIAL REQUIRED FOR ROADS, STREETS AND ALLEYS
1—2—4 MIX.

Quantity in Barrels of Cement—Cubic Yards of Sand—Cubic Yards of Stone.

Cement Required:—1.48 Barrels Per Cubic Yard.
Sand Required:—.45 Cubic Yards Per Cubic Yard.
Stone Required:—.90 Cubic Yards Per Cubic Yard.

Thickness in Inches			Width in Feet											
			20		22		24		26		28		30	
Edge	Center	Average	Foot	Mile	Foot	Mile	Foot	Mile	Foot	Mile	Foot	Mile	Foot	Mile
6	8	7½												
	CEMENT		.669	3533	.737	3891	.804	4243	.870	4594	.937	4946	1.005	5306
	SAND		.20	1074	.22	1183	.24	1290	.26	1397	.28	1504	.31	1613
	STONE		.41	2148	.45	2366	.49	2580	.53	2794	.57	3008	.61	3227
6	9	8												
	CEMENT		.731	3860	.804	4243	.878	4634	.950	5017	1.023	5399	1.097	5790
	SAND		.22	1174	.24	1290	.27	1409	.29	1526	.31	1642	.33	1760
	STONE		.44	2347	.49	2580	.53	2818	.58	3051	.62	3283	.67	3521
7	7	7												
	CEMENT		.639	3376	.703	3712	.767	4048	.832	4391	.895	4727	.959	5063
	SAND		.19	1026	.21	1129	.23	1231	.25	1335	.27	1437	.29	1539
	STONE		.39	2053	.43	2257	.47	2462	.51	2670	.54	2875	.58	3079
7	8	7½												
	CEMENT		.700	3696	.770	4064	.839	4431	.910	4806	.980	5173	1.049	5541
	SAND		.21	1124	.23	1236	.26	1347	.28	1461	.30	1573	.32	1685
	STONE		.43	2247	.47	2471	.51	2695	.55	2922	.60	3146	.64	3370
7	9	8½												
	CEMENT		.761	4017	.838	4422	.913	4822	.989	5220	1.066	5627	1.133	6025
	SAND		.23	1221	.25	1345	.28	1466	.30	1587	.32	1711	.35	1831
	STONE		.46	2443	.51	2689	.56	2932	.60	3174	.65	3422	.69	3664
7	10	9												
	CEMENT		.821	4336	.904	4774	.987	5213	1.069	5642	1.151	6080	1.233	6509
	SAND		.25	1319	.27	1452	.30	1585	.32	1715	.35	1849	.37	1979
	STONE		.50	2637	.55	2903	.60	3170	.65	3431	.70	3697	.75	3958
8	8	8												
	CEMENT		.731	3860	.804	4243	.878	4634	.950	5017	1.023	5399	1.097	5790
	SAND		.22	1174	.24	1290	.27	1409	.29	1526	.31	1642	.33	1760
	STONE		.44	2347	.49	2580	.53	2818	.58	3051	.62	3283	.67	3521
8	9	8½												
	CEMENT		.792	4181	.870	4594	.949	5008	1.029	5432	1.107	5845	1.187	6268
	SAND		.24	1271	.26	1397	.29	1523	.31	1652	.34	1777	.36	1906
	STONE		.48	2543	.53	2794	.58	3046	.63	3303	67	3554	.72	3812
8	10	9½												
	CEMENT		.851	4493	.938	4955	1.023	5399	1.107	5845	1.193	6299	1.279	6752
	SAND		.26	1366	.29	1507	.31	1642	.34	1777	.36	1915	.39	2053
	STONE		.52	2732	.57	3013	.62	3283	.67	3554	.73	3830	.78	4106

Table No. 11

QUANTITY OF MATERIAL REQUIRED FOR ROADS, STREETS AND ALLEYS

1—2½—5 MIX.

Quantity in Barrels of Cement—Cubic Yards of Sand—Cubic Yards of Stone.

Cement Required:—1.21 Barrels Per Cubic Yard
Sand Required:—.46 Cubic Yards Per Cubic Yard.
Stone Required:—.92 Cubic Yards Per Cubic Yard.

Thickness in Inches				Width in Feet											
				9		10		12		14		16		18	
Edge	Center	Average		Foot	Mile	Foot	Mile	Foot	Mile	Foot	Mile	Foot	Mile	Foot	Mile
4	4	4													
			CEMENT	134	709	149	785	.179	945	209	1105	.238	1258	.269	1418
			SAND	.05	270	.06	299	.07	359	.08	420	.09	478	.10	539
			STONE	.10	539	.11	597	.14	719	16	840	.18	957	.20	1078
4	5	4⅔													
			CEMENT	.157	830	.174	920	.209	1105	.243	1284	.278	1469	.313	1655
			SAND	.06	316	.07	350	.08	420	.09	488	.11	558	.12	629
			STONE	12	631	.13	699	16	840	18	976	.21	1117	.24	1259
4	6	5⅓													
			CEMENT	.179	945	.198	1048	.238	1258	.278	1469	.318	1681	.358	1891
			SAND	.07	359	.08	398	.09	478	.11	558	.12	639	.14	719
			STONE	.14	719	.15	797	.18	957	.21	1117	.24	1278	.27	1438
5	5	5													
			CEMENT	.168	888	.186	984	.224	1182	.261	1379	.299	1578	.336	1776
			SAND	06	338	.07	374	.09	449	.10	524	.11	600	.13	675
			STONE	.13	675	.14	748	.17	899	.20	1049	.23	1200	.26	1351
5	6	5⅔													
			CEMENT	.190	1003	.212	1118	.254	1342	.296	1566	.339	1788	.380	2006
			SAND	.07	381	.08	425	.10	510	.11	595	.13	680	.14	763
			STONE	.14	763	.16	850	.19	1020	.23	1190	.26	1360	.29	1525
5	7	6⅓													
			CEMENT	.212	1124	.236	1246	.283	1496	.332	1751	.379	2000	.426	2249
			SAND	.08	427	.09	474	.11	569	.13	666	.14	760	.16	855
			STONE	.16	855	.18	948	.22	1137	.25	1331	.29	1521	.32	1710
5	8	7													
			CEMENT	.235	1239	.261	1379	.313	1655	.365	1930	.419	2211	.471	2485
			SAND	.09	471	.10	524	.12	629	.14	734	.16	840	.18	945
			STONE	.18	942	.20	1049	.24	1259	.28	1467	.32	1681	.36	1890
6	6	6													
			CEMENT	.202	1067	.224	1182	.269	1418	.313	1655	.358	1891	.403	2127
			SAND	.08	406	.09	449	.10	539	.12	629	.14	719	.15	809
			STONE	.15	811	.17	899	.20	1078	.24	1259	.27	1438	.31	1617
6	7	6⅔													
			CEMENT	.224	1182	.248	1309	.299	1578	.348	1840	.398	2102	.448	2364
			SAND	.09	449	.09	498	.11	600	.13	700	.15	799	.17	899
			STONE	.17	899	.19	995	.23	1200	.26	1400	.30	1598	.34	1798

Table No. 11
QUANTITY OF MATERIAL REQUIRED FOR ROADS, STREETS AND ALLEYS
1—2½—5 MIX.

Quantity in Barrels of Cement—Cubic Yards of Sand—Cubic Yards of Stone.

Cement Required:—1.21 Barrels Per Cubic Yard.
Sand Required:—.46 Cubic Yards Per Cubic Yard.
Stone Required:—.92 Cubic Yards Per Cubic Yard.

Thickness in Inches (Edge / Center / Average)		Width in Feet											
		20		22		24		26		28		30	
		Foot	Mile	Foot	Mile	Foot	Mile	Foot	Mile	Foot	Mile	Foot	Mile
4 4 4	CEMENT	.299	1578	.328	1732	.358	1891	.388	2051	.419	2211	.448	2364
	SAND	.11	600	.12	658	.14	719	.15	780	.16	840	.17	899
	STONE	.23	1200	.25	1317	.27	1438	.30	1559	.32	1681	.34	1798
4 5 4¾	CEMENT	.348	1840	.382	2018	.417	2205	.453	2390	.488	2575	.512	2754
	SAND	.13	700	.15	767	.16	838	.17	909	.19	979	.20	1047
	STONE	.26	1400	.29	1535	.32	1676	.34	1817	.37	1958	.40	2094
4 6 5⅓	CEMENT	.398	2102	.438	2312	.478	2524	.518	2735	.558	2945	.598	3156
	SAND	.15	799	.17	879	.18	960	.20	1040	.21	1120	.23	1200
	STONE	.30	1598	.33	1758	.36	1919	.39	2079	.42	2239	.45	2399
5 5 5	CEMENT	.374	1975	.410	2166	.448	2364	.485	2562	.523	2760	.560	2958
	SAND	.14	751	.16	823	.17	899	.18	974	.20	1049	.21	1125
	STONE	.28	1501	.31	1647	.34	1798	.37	1948	.40	2099	.43	2249
5 6 5⅓	CEMENT	.422	2230	.465	2454	.507	2677	.549	2900	.592	3124	.634	3348
	SAND	.16	848	.18	933	.19	1018	.21	1103	.22	1188	.24	1273
	STONE	.32	1696	.35	1866	.39	2035	.42	2205	.45	2375	.48	2546
5 7 6⅓	CEMENT	.473	2497	.520	2747	.567	2996	.615	3245	.662	3494	.709	3744
	SAND	.18	949	.20	1044	.22	1139	.23	1234	.25	1328	.27	1423
	STONE	.36	1899	.40	2088	.43	2278	.47	2467	.50	2657	.54	2846
5 8 7	CEMENT	.523	2760	.575	3035	.627	3309	.680	3590	.732	3865	.784	4139
	SAND	.20	1049	.22	1154	.24	1258	.26	1365	.28	1469	.30	1574
	STONE	.40	2099	.44	2307	.48	2516	.52	2730	.56	2938	.60	3147
6 6 6	CEMENT	.448	2364	.492	2600	.537	2836	.582	3073	.628	3315	.673	3553
	SAND	.17	899	.19	989	.20	1078	.22	1168	.24	1260	.26	1351
	STONE	.34	1798	.37	1977	.41	2156	.44	2337	.48	2521	.51	2701
6 7 6⅓	CEMENT	.497	2626	.547	2888	.597	3150	.646	3412	.697	3680	.747	3942
	SAND	.19	998	.21	1098	.23	1197	.25	1297	.26	1399	.28	1499
	STONE	.38	1996	.42	2196	.45	2395	.49	2594	.53	2798	.57	2997

Table No. 11
QUANTITY OF MATERIAL REQUIRED FOR ROADS, STREETS AND ALLEYS
1—2½—5 MIX.

Quantity in Barrels of Cement—Cubic Yards of Sand—Cubic Yards of Stone.

Cement Required:—1.21 Barrels Per Cubic Yard.
Sand Required:—.46 Cubic Yards Per Cubic Yard.
Stone Required:—.92 Cubic Yards Per Cubic Yard.

Thickness in Inches (Edge / Center / Average)			9		10		12		14		16		18	
Edge	Center	Average	Foot	Mile	Foot	Mile	Foot	Mile	Foot	Mile	Foot	Mile	Foot	Mile
6	8	7½												
CEMENT			.247	1303	.273	1444	.328	1732	.384	2026	.438	2312	.492	2600
SAND			.09	495	.10	549	.12	658	.15	770	.17	879	.19	989
STONE			.19	991	.21	1098	.25	1317	.29	1540	.33	1758	.37	1977
6	9	8												
CEMENT			.269	1418	.299	1578	.358	1891	.419	2211	.478	2524	.537	2836
SAND			.10	539	.11	600	.14	719	.16	840	.18	960	.20	1078
STONE			.20	1078	.23	1200	.27	1438	.32	1681	.36	1919	.41	2156
7	7	7												
CEMENT			.235	1239	.261	1379	.313	1655	.365	1930	.419	2211	.471	2485
SAND			.09	471	.10	524	.12	629	.14	734	.16	840	.18	945
STONE			.18	942	.20	1049	.24	1259	.28	1467	.32	1681	.36	1890
7	8	7⅔												
CEMENT			.258	1361	.286	1508	.344	1815	.401	2115	.457	2415	.515	2721
SAND			.10	518	.11	573	.13	690	.15	804	.17	918	.20	1035
STONE			.20	1035	.22	1146	.26	1380	.30	1608	.35	1836	.39	2069
7	9	8½												
CEMENT			.280	1476	.311	1642	.374	1975	.436	2300	.497	2626	.560	2958
SAND			.11	561	.12	624	.14	751	.17	874	.19	998	.21	1125
STONE			.21	1122	.24	1248	.28	1501	.33	1749	.38	1996	.43	2349
7	10	9												
CEMENT			.303	1597	.336	1776	.403	2127	.471	2485	.537	2836	.605	3194
SAND			.12	607	.13	675	.15	809	.18	945	.20	1078	.23	1214
STONE			.23	1214	.26	1351	.31	1617	.36	1890	.41	2156	.46	2429
8	8	8												
CEMENT			.269	1418	.299	1578	.358	1891	.419	2211	.478	2524	.537	2836
SAND			.10	539	.11	600	.14	719	.16	840	.18	960	.20	1078
STONE			.20	1078	.23	1200	.27	1438	.32	1681	.36	1919	.41	2156
8	9	8⅔												
CEMENT			.292	1539	.323	1706	.388	2051	.453	2390	.518	2735	.582	3073
SAND			.11	585	.12	649	.15	780	.17	909	.20	1040	.22	1168
STONE			.22	1170	.25	1297	.30	1559	.34	1817	.39	2079	.44	2337
8	10	9½												
CEMENT			.313	1655	.348	1840	.419	2211	.488	2575	.558	2945	.627	3309
SAND			.12	629	.13	700	.16	840	.19	979	.21	1120	.24	1258
STONE			.24	1259	.26	1400	.32	1681	.37	1958	.42	2239	.48	2516

Table No. 11
QUANTITY OF MATERIAL REQUIRED FOR ROADS, STREETS AND ALLEYS
1—2½—5 MIX.

Quantity in Barrels of Cement—Cubic Yards of Sand—Cubic Yards of Stone.

Cement Required:—1.21 Barrels Per Cubic Yard.
Sand Required:—.46 Cubic Yards Per Cubic Yard.
Stone Required:—.92 Cubic Yards Per Cubic Yard.

Thickness in Inches			Width in Feet											
Edge	Center	Average	20		22		24		26		28		30	
			Foot	Mile	Foot	Mile	Foot	Mile	Foot	Mile	Foot	Mile	Foot	Mile
6	8	7½												
	CEMENT		.547	2888	.603	3181	.657	3469	.711	3757	.766	4044	.822	4338
	SAND		.21	1098	.23	1209	.25	1319	.27	1428	.29	1537	.31	1649
	STONE		.42	2196	.46	2419	.50	2638	.54	2857	.58	3075	.62	3298
6	9	8												
	CEMENT		.598	3156	.657	3469	.718	3789	.777	4102	.836	4414	.897	4734
	SAND		.23	1200	.25	1319	.27	1440	.30	1559	.32	1678	.34	1800
	STONE		.45	2399	.50	2638	.55	2881	.59	3119	.64	3356	.68	3599
7	7	7												
	CEMENT		.523	2760	.575	3035	.627	3309	.680	3590	.732	3865	.784	4139
	SAND		.20	1049	.22	1154	.24	1258	.26	1365	.28	1469	.30	1574
	STONE		.40	2099	.44	2307	.48	2516	.52	2730	.56	2938	.60	3147
7	8	7½												
	CEMENT		.572	3021	.629	3323	.686	3623	.744	3929	.801	4229	.858	4530
	SAND		.22	1149	.24	1263	.26	1377	.28	1494	.30	1608	.33	1722
	STONE		.44	2297	.48	2526	.52	2754	.57	2987	.61	3215	.65	3444
7	9	8½												
	CEMENT		.622	3284	.685	3615	.747	3942	.808	4268	.871	4600	.933	4926
	SAND		.24	1248	.26	1374	.28	1499	.30	1622	.33	1749	.35	1873
	STONE		.47	2497	.52	2749	.57	2997	.61	3245	.66	3498	.71	3745
7	10	9												
	CEMENT		.672	3545	.739	3903	.807	4262	.874	4613	.941	4971	1.008	5322
	SAND		.26	1348	.28	1484	.31	1620	.33	1754	.36	1890	.38	2023
	STONE		.51	2696	.56	2978	.61	3240	.66	3507	.72	3779	.77	4046
8	8	8												
	CEMENT		.598	3156	.657	3469	.718	3789	.777	4102	.836	4414	.897	4734
	SAND		.23	1200	.25	1319	.27	1440	.30	1559	.32	1678	.34	1800
	STONE		.45	2399	.50	2638	.55	2881	.59	3119	.64	3356	.68	3599
8	9	8½												
	CEMENT		.647	3418	.711	3757	.776	4095	.841	4441	.905	4778	.970	5124
	SAND		.25	1300	.27	1428	.29	1557	.32	1688	.34	1817	.37	1948
	STONE		.49	2599	.54	2857	.59	3113	.64	3376	.69	3633	.74	3896
8	10	9½												
	CEMENT		.696	3674	767	4051	.836	4414	.905	4778	.975	5150	1.045	5520
	SAND		.26	1397	.29	1540	.32	1678	.34	1817	.37	1958	.40	2099
	STONE		.53	2793	.58	3080	.64	3356	.69	3633	.74	3916	.79	4197

Table No. 11
QUANTITY OF MATERIAL REQUIRED FOR ROADS, STREETS AND ALLEYS
1—3—6 MIX.

Quantity in Barrels of Cement—Cubic Yards of Sand—Cubic Yards of Stone.

Cement Required:—1.02 Barrels Per Cubic Yard.
Sand Required:—.47 Cubic Yards Per Cubic Yard.
Stone Required:—.93 Cubic Yards Per Cubic Yard.

Thickness in Inches (Edge / Center / Average)			Width in Feet											
			9		10		12		14		16		18	
Edge	Center	Average	Foot	Mile	Foot	Mile	Foot	Mile	Foot	Mile	Foot	Mile	Foot	Mile
6	8	7½												
CEMENT			.208	1099	.231	1217	.276	1460	.323	1707	.369	1949	.415	2192
SAND			.10	506	.11	561	.13	673	.15	787	.17	898	.19	1010
STONE			.19	1002	.21	1109	.25	1331	.29	1557	.34	1777	.38	1999
6	9	8												
CEMENT			.226	1195	.252	1330	.302	1594	.353	1864	.403	2128	.452	2391
SAND			.10	551	.12	613	.14	735	.16	859	.19	980	.21	1102
STONE			.21	1090	.23	1213	.28	1454	.32	1699	.37	1940	.41	2180
7	7	7												
CEMENT			.198	1044	.220	1163	.264	1395	.308	1627	.353	1864	.397	2095
SAND			.09	481	.10	536	.12	629	.14	750	.16	859	.18	965
STONE			.18	952	.20	1060	.24	1272	.28	1483	.32	1699	.36	1910
7	8	7¾												
CEMENT			.217	1146	.241	1271	.290	1530	.338	1783	.386	2036	.435	2294
SAND			.10	528	.11	586	.13	705	.16	822	.18	938	.20	1057
STONE			.20	1045	.22	1159	.26	1395	.31	1626	.35	1856	.40	2092
7	9	8½												
CEMENT			.236	1244	.262	1384	.315	1665	.367	1939	.419	2213	.472	2494
SAND			.11	573	.12	638	.15	767	.17	893	.19	1020	.22	1149
STONE			.21	1135	.24	1262	.29	1518	.33	1768	.38	2018	.43	2274
7	10	9												
CEMENT			.255	1346	.284	1497	.340	1793	.397	2095	.453	2391	.510	2693
SAND			.12	620	.13	690	.16	826	.18	965	.21	1102	.24	1241
STONE			.23	1228	.26	1365	.31	1635	.36	1910	.42	2180	.47	2455
8	8	8												
CEMENT			.226	1195	.252	1330	.302	1594	.353	1864	.403	2128	.452	2391
SAND			.10	551	.12	613	.14	735	.16	859	.19	980	.21	1102
STONE			.21	1090	.23	1213	.28	1454	.32	1699	.37	1940	.41	2180
8	9	8½												
CEMENT			.246	1297	.272	1438	.327	1729	.381	2015	.437	2305	.491	2591
SAND			.11	598	.13	663	.15	797	.18	928	.20	1062	.23	1194
STONE			.22	1183	.25	1311	.30	1576	.35	1837	.40	2102	.45	2362
8	10	9½												
CEMENT			.264	1395	.294	1551	.353	1864	.411	2171	.470	2483	.528	2790
SAND			.12	643	.14	715	.16	859	.19	1000	.22	1144	.24	1285
STONE			.24	1272	.27	1415	.32	1699	.37	1979	.43	2264	.48	2544

Table No. 11

QUANTITY OF MATERIAL REQUIRED FOR ROADS, STREETS AND ALLEYS

1—3—6 MIX.

Quantity in Barrels of Cement—Cubic Yards of Sand—Cubic Yards of Stone.

Cement Required:—1.02 Barrels Per Cubic Yard.
Sand Required:—.47 Cubic Yards Per Cubic Yard.
Stone Required:—.93 Cubic Yards Per Cubic Yard.

| Thickness in Inches | | | Width in Feet | | | | | | | | | | | |
| Edge | Center | Average | 20 | | 22 | | 24 | | 26 | | 28 | | 30 | |
			Foot	Mile	Foot	Mile	Foot	Mile	Foot	Mile	Foot	Mile	Foot	Mile
6	8	7½												
CEMENT			.461	2435	.508	2682	.554	2924	.600	3167	.646	3409	.693	3657
SAND			.21	1122	.23	1236	.26	1349	.28	1459	.30	1571	.32	1685
STONE			.42	2220	.46	2445	.50	2666	.55	2888	.59	3108	.63	3334
6	9	8												
CEMENT			.504	2660	.554	2924	.605	3194	.655	3458	.705	3721	.756	3990
SAND			.23	1226	.26	1347	.28	1472	.30	1593	.32	1715	.35	1839
STONE			.46	2425	.50	2666	.55	2912	.60	3153	.64	3393	.69	3638
7	7	7												
CEMENT			.441	2327	.485	2558	.528	2790	.573	3026	.617	3259	.661	3489
SAND			.20	1072	.22	1179	.24	1285	.26	1394	.28	1501	.30	1608
STONE			.40	2121	.44	2332	.48	2544	.52	2759	.56	2970	.60	3182
7	8	7½												
CEMENT			.482	2547	.530	2801	.578	3054	.627	3312	.675	3565	.723	3819
SAND			.22	1174	.24	1291	.27	1407	.29	1526	.31	1643	.33	1760
STONE			.44	2322	.48	2554	.53	2784	.57	3020	.62	3250	.66	3482
7	9	8½												
CEMENT			.524	2768	.577	3048	.629	3323	.681	3598	.734	3878	.786	4152
SAND			.24	1276	.27	1404	.29	1531	.31	1658	.34	1787	.362	1913
STONE			.48	2524	.53	2779	.57	3030	.62	3280	.67	3536	.72	3786
7	10	9												
CEMENT			.566	2989	.623	3291	.680	3592	.736	3888	.794	4190	.850	4486
SAND			.26	1377	.29	1516	.31	1655	.34	1792	.37	1931	.39	2067
STONE			.52	2725	.57	3000	.62	3275	.67	3545	.72	3820	.77	4090
8	8	8												
CEMENT			.504	2660	.554	2924	.605	3194	.655	3458	.705	3721	.756	3990
SAND			.23	1226	.26	1347	.28	1472	.30	1593	.32	1715	.35	1839
STONE			.46	2425	.50	2666	.55	2912	.60	3153	.64	3393	.69	3638
8	9	8½												
CEMENT			.546	2882	.600	3167	.654	3452	.709	3743	.763	4028	.818	4320
SAND			.25	1328	.28	1459	.30	1590	.33	1725	.35	1856	.38	1990
STONE			.50	2627	.55	2888	.60	3147	.65	3413	.70	3673	.75	3939
8	10	9½												
CEMENT			.587	3097	.647	3415	.705	3721	.763	4028	.822	4341	.881	4653
SAND			.27	1427	.30	1574	.32	1715	.35	1856	.38	2000	.41	2144
STONE			.53	2823	.59	3114	.64	3393	.70	3673	.75	3958	.80	4243

Table No. 11

QUANTITY OF MATERIAL REQUIRED FOR ROADS, STREETS AND ALLEYS

1—3—6 MIX.

Quantity in Barrels of Cement—Cubic Yards of Sand—Cubic Yards of Stone.

Cement Required:—1.02 Barrels Per Cubic Yard.
Sand Required:—.47 Cubic Yards Per Cubic Yard.
Stone Required:—.93 Cubic Yards Per Cubic Yard.

Thickness in Inches				Width in Feet											
				9		10		12		14		16		18	
Edge	Center	Average		Foot	Mile	Foot	Mile	Foot	Mile	Foot	Mile	Foot	Mile	Foot	Mile
4	4	4	CEMENT	.113	598	.125	662	.151	797	.176	931	.200	1061	.226	1195
			SAND	.05	275	.06	305	.07	367	.08	429	.09	489	.10	551
			STONE	.10	545	.11	604	.14	726	.15	849	.18	967	.21	1090
4	5	4½	CEMENT	.133	700	.147	775	.176	931	.205	1082	.235	1238	.264	1395
			SAND	.06	322	.07	357	.08	429	.09	499	.10	571	.12	643
			STONE	.12	638	.13	707	.16	849	.19	987	.21	1129	.24	1272
4	6	5½	CEMENT	.151	797	.167	883	.201	1061	.235	1238	.268	1417	.302	1594
			SAND	.07	367	.08	407	.09	489	.11	571	.12	653	.14	735
			STONE	.14	726	.15	805	.18	967	.21	1129	.24	1292	.28	1454
5	5	5	CEMENT	.142	749	.157	829	.189	997	.220	1163	.252	1330	.284	1497
			SAND	.07	345	.07	382	.09	459	.10	536	.12	613	.13	690
			STONE	.13	683	.14	756	.17	909	.20	1060	.23	1213	.26	1365
5	6	5½	CEMENT	.160	846	.179	942	.214	1131	.250	1320	.286	1508	.320	1691
			SAND	.07	390	.08	434	.10	521	.12	608	.13	695	.15	779
			STONE	.15	771	.16	859	.20	1031	.23	1203	.26	1375	.29	1542
5	7	6½	CEMENT	.180	948	.199	1051	.239	1261	.279	1476	.319	1686	.359	1896
			SAND	.08	437	.09	484	.11	581	.13	681	.15	777	.17	871
			STONE	.16	864	.18	958	.22	1149	.25	1346	.29	1537	.32	1729
5	8	7	CEMENT	.198	1044	.220	1163	.264	1395	.308	1627	.353	1864	.397	2095
			SAND	.09	481	.10	536	.12	643	.14	750	.16	859	.18	965
			STONE	.18	952	.20	1060	.24	1272	.28	1483	.32	1699	.36	1910
6	6	6	CEMENT	.170	900	.189	997	.226	1195	.264	1395	.302	1594	.340	1793
			SAND	.08	415	.09	459	.10	551	.12	643	.14	735	.16	826
			STONE	.16	820	.17	909	.21	1090	.24	1272	.28	1454	.31	1635
6	7	6½	CEMENT	.189	997	.209	1104	.252	1330	.294	1551	.336	1772	.377	1993
			SAND	.09	459	.10	509	.12	613	.14	715	.15	816	.17	918
			STONE	.17	909	.19	1006	.23	1213	.27	1415	.31	1615	.34	1817

Table No. 11
QUANTITY OF MATERIAL REQUIRED FOR ROADS, STREETS AND ALLEYS
1—3—6 MIX.
Quantity in Barrels of Cement—Cubic Yards of Sand—Cubic Yards of Stone.

Cement Required:—1.02 Barrels Per Cubic Yard.
Sand Required:—.47 Cubic Yards Per Cubic Yard.
Stone Required:—.93 Cubic Yards Per Cubic Yard.

Thickness in Inches			Width in Feet										
Edge	Center	Average	20		22		24		26		28		30	
			Foot	Mile	Foot	Mile	Foot	Mile	Foot	Mile	Foot	Mile	Foot	Mile
4	4	4												
	CEMENT		.252	1330	.276	1460	.302	1594	.327	1729	.353	1864	.377	1993
	SAND		.12	613	.13	673	.14	735	.15	797	.16	859	.17	918
	STONE		.23	1213	.25	1331	.28	1454	.30	1576	.32	1699	.34	1817
4	5	4½												
	CEMENT		.294	1551	.322	1701	.352	1858	.381	2015	.411	2171	.440	2322
	SAND		.14	715	.15	784	.16	856	.18	928	.19	1000	.20	1070
	STONE		.27	1415	.29	1551	.32	1694	.35	1837	.37	1979	.40	2117
4	6	5½												
	CEMENT		.336	1772	.369	1949	.403	2128	.437	2305	.470	2483	.504	2660
	SAND		.15	816	.17	898	.19	980	.20	1062	.22	1144	.23	1226
	STONE		.31	1615	.34	1777	.37	1940	.40	2102	.43	2264	.46	2425
5	5	5												
	CEMENT		.315	1665	.346	1826	.377	1993	.409	2159	.441	2327	.472	2494
	SAND		.15	767	.16	841	.17	918	.19	995	.20	1072	.22	1149
	STONE		.29	1518	.32	1665	.34	1817	.37	1968	.40	2121	.43	2274
5	6	5½												
	CEMENT		.360	1880	.392	2069	.427	2256	.463	2445	.499	2644	.534	2822
	SAND		.16	866	.18	953	.20	1040	.21	1127	.23	1214	.25	1300
	STONE		.32	1714	.36	1886	.39	2057	.42	2229	.45	2401	.49	2573
5	7	6½												
	CEMENT		.399	2105	.439	2315	.478	2526	.518	2736	.558	2946	.598	3156
	SAND		.18	970	.20	1067	.22	1164	.24	1261	.26	1357	.28	1454
	STONE		.36	1920	.40	2111	.44	2303	.47	2494	.51	2686	.54	2877
5	8	7												
	CEMENT		.441	2327	.485	2558	.528	2790	.573	3026	.617	3259	.661	3489
	SAND		.20	1072	.22	1179	.24	1285	.26	1394	.28	1501	.30	1608
	STONE		.40	2121	.44	2332	.48	2544	.52	2759	.56	2970	.60	3182
6	6	6												
	CEMENT		.377	1993	.415	2192	.453	2391	.491	2591	.529	2795	.567	2995
	SAND		.17	918	.19	1010	.21	1102	.23	1194	.24	1288	.26	1380
	STONE		.34	1817	.37	1999	.41	2180	.45	2362	.48	2548	.52	2730
6	7	6½												
	CEMENT		.419	2213	.461	2435	.503	2655	.545	2876	.588	3102	.629	3323
	SAND		.19	1020	.21	1122	.23	1223	.25	1325	.27	1429	.29	1533
	STONE		.38	2018	.42	2220	.46	2421	.50	2622	.54	2828	.57	3030

The Koehring Dandie Mixer.
Equipped with power charging skip and solid rubber tires.

CHAPTER 4.

MISCELLANEOUS NOTES FOR SUPERIN-TENDENT AND FOREMAN

Forms

Various methods of building and erecting forms, and other important essentials related to their use, have been more fully discussed elsewhere. There are, however, a number of precautions which should always be observed which are of such importance that repetition of mention can be excused.

Forms should be built exactly as called for by the drawings unless proved impractical. The superintendent should be able to determine whether forms are sufficiently strong to support the load of concrete that will be placed upon them and should see that they are sufficiently braced so that they will neither collapse nor sag when filled. They should be cleaned of all refuse or rubbish before any concrete is placed. This is particularly true of column forms, but special reference to column form does not mean that any part of the inspection is to be slighted.

Reinforcing steel must be properly placed and when in place should be carefully checked against the drawings to make certain that the required amount has been used and is in proper position.

Many failures have been caused by weakness of supports for concrete forms and centering. If the forms shake or vibrate considerably when work is in progress, it is almost certain that such disturbance will affect the concrete while hardening. An interference of this kind during the hardening of the concrete impairs the effectiveness of the cement as a binder.

Forms should be so placed and so supported, particularly when the uprights rest upon the earth, as to prevent warping, twisting or sagging. The maximum safe loads for wood posts of various lengths and sections are given below. Knowing the length of post, total weight of concrete and form to be supported and the economical number of posts, the load per post and size can readily be determined.

A corner post carries only one-half of the load carried by the side post, while a side post carries one-half the load that must be borne by the inside one.

Table No. 12
MAXIMUM SAFE LOAD FOR WOOD COLUMNS

Length in Feet	Minimum Dimensions		
	4 in.	6 in.	8 in.
5	9,400		
6	8,800		
7	8,200		
8	7,500	20,700	
9	6,900	19,800	
10	6,300	18,900	37,700
11		17,900	36,400
12		17,000	35,200
14		15,100	32,700
16			30,200
18			27,600
20			25,100

Example:—Flat slab 14 feet by 17 feet 8 inches, weighing approximately 60,000 pounds, 16 posts can be spaced economically in four rows of 4 each. There will be 4 corner posts, 8 side posts and 4 inside posts—16 posts.

4 corner posts carry load of 1 inside post=1
8 side posts carry load of 4 inside posts=4
4 inside posts carry load of 4 inside posts=4

Number of posts of equal load 9

Maximum load per post$=\dfrac{60,000}{9}=$6,666 lbs.

Length of post 6 feet 0 inches.

From the table we find one 4 by 4 post 6 feet long will carry safely a load of 8,800 pounds, since posts should never be less than 4 by 4 inches and this size will answer in this case.

In filling forms, care should be taken to place the proper quantities of concrete at one time. It should be placed in layers no thicker or deeper than can be properly consolidated and caused to unite with concrete previously placed. The forms for a slab and beam should be filled at the same time. If necessary to discontinue work, good judgment should be exercised as to the best place to make such a stop so as not to leave a permanent line of cleavage that will affect the final strength of the structure.

Careful Supervision over Proportioning and Mixing Necessary

At all times while concreting is in progress, the contractor's superintendent or foreman should keep careful watch over the manner in which the concrete is being proportioned and mixed, so that there will be no question but that specification requirements are being complied with. Cement is sold by the barrel but is usually received on the job in bags or sacks, four of such containers corresponding to a so-called barrel. At all times careful watch should be kept of the aggregates being used to make certain that definite proportions called for are not changed, due to carelessness of some of the workmen, by being carelessly measured, nor that unscreened aggregates are being substituted for the separate volumes of fine and course aggregates specified.

In no instance should forms or centering be removed until it is positively known that the concrete has hardened sufficiently to have the required strength, not only to carry its own weight but any weight that may be placed on it during subsequent processes of construction. Concrete hardens much more rapidly during warm than in cold, damp weather. Retaining walls that are to withstand earth pressure should not be subjected to such pressure until all possibility of injury from load has passed. There is no good guide for form removal other than that acquired by long experience, which enables the superintendent or foreman to place proper value upon the conditions under which the concrete has been hardening. Extraordinary precautions should be taken when it comes to removing supports from floor slabs, roofs, arches and similar classes of work. The length of time that forms are left in place should be much longer in cold than in warm weather. Somewhat of a guide can be obtained by making cubes or cylinders of concrete at the same time as the concrete in the job is placed. These should be examined later and tested, if necessary, to determine whether the concrete is of proper strength. Yet such tests must not be relied upon as an invariable guide.

Concreting in Cold Weather

Experience has proved the possibility of carrying on many classes of concrete work under conditions which a few years ago would have been thought unfavorable. There are contracting firms who specialize on concrete work in cold weather. Such lengthening of the ordinary concreting season has been made possible by the application of such precautions as heating the sand, pebbles or broken stone and water so that sufficient warmth is added to the concrete mixture to carry it through the period necessary for early hardening. All materials except the cement are heated so that the concrete when placed has a temperature of from seventy to eighty degrees Fahrenheit. After being placed, it is protected so that the mass will retain this heat and the possibility of freezing be prevented for at least forty-eight hours. Work so done will be as successful as that carried on under the usual favorable conditions. This is true of practically all classes of concrete work except concrete road or street construction, which it is not advisable to continue when temperatures fall so low that continued freezing may be expected.

No concrete should be laid when temperatures may go to freezing or lower, unless precautions are taken to heat materials as described. It is better to suspend work than to run the chance of a severe drop in temperature which may injure the concrete if it is frozen more than once before thoroughly hardened. Concrete floors are laid in buildings during freezing weather by enclosing the frame of the structure with with tarpaulins and maintaining heat within the enclosure by means of stoves, or "salamanders".

In heating materials, a number of methods may be employed. Mixing water can readily be heated by leading the water through a pipe coil around which a coke or wood fire is kept. Or it can be heated in barrels by discharging steam from the boiler plant on the job into the barrels filled with water; or if there is no such plant, then the required quantities can usually be heated in large kettles. The sand and pebbles or broken stone can be heated by improvising pipe stoves made out of sections of metal smoke stack and piling the materials over and about a piece of such stack with fire built inside. Materials should be turned or raked over frequently so those next to the stove will not be damaged by over-heating, and those farther away will be sure to have the frost drawn from them. Another method frequently employed to heat the stone or sand is to insert a small perforated pipe through which live steam is passed. It is absolutely necessary to get all frost out of the aggregate before putting it in a concrete mixture.

Very often on account of the limited space in the street or at the site of the work, or because of unusual conditions, it is not easy to improvise the desired method of heating material. A heating attachment which was originally designed for use in mixing asphaltic concrete, has been very sucessfully used to mix concrete in winter where it is desirable to have all materials thoroughly heated. The heating unit consists of an oil tank for supplying fuel to the double burner oil furnace and a blower which forces the heated air and flame into the inside of the drum, thus insuring the necessary degree of warmth to the concrete which, with other measures of protection, prevent it from freezing in the forms until after early hardening has been completed.

Various methods are used to protect the concrete after it has been placed, from possible damage due to freezing. Coverings of canvas, straw or similar material can be used on flat surfaces, while vertical faces also may be covered with canvas hung away from the forms; or the forms can be battened and building paper tacked on the battens, thus introducing a dead space that does much to insulate against extreme cold. For mass work sometimes the forms themselves afford sufficient protection when cold is not extreme. Padding the forms with hay or straw in extreme cold weather provides excellent protection.

Manure should never be placed directly on concrete as a means of protection against freezing. Experiments have proved that in the process of decomposition the chemical changes which may take place in the manure are likely to produce nitric acid. Although this does not always happen, it may, and if it does the result will be a scaling or pitting of the surface. Besides, manure placed directly in contact with a concrete walk or floor is certain to cause considerable staining of the surface, which in most cases would be objectionable.

Curing of Concrete

Many persons have the impression that the hardening process which takes place in concrete is due to drying. Nothing will do more to weaken concrete than to allow it to dry out rapidly after placed. For this reason concrete work requires certain protection in warm weather to safeguard it, just as it requires protection in cold weather. When sun or hot dry winds strike fresh concrete much of the water introduced in mixing rapidly evaporates. Water is necessary to the hardening of concrete. Generally speaking, the protection that should be given to concrete during hot weather is essentially the same for all classes of concrete work. Ways and means of applying the protection may differ slightly, but all aim toward the same end.

Pavements in general, which properly include floors, sidewalks, driveways, etc., have a relatively larger area exposed to the atmosphere than has mass concrete. Wall sections have a still greater area exposed, usually two sides. Either canvas or earth covering should be applied to concrete surfaces to protect against rapid drying. In hot weather it is very desirable to stretch canvas on frames over concrete street and highway pavement immediately after the surface has been struck off and floated, and to keep such a covering in place until the surface has hardened enough to permit applying a protective layer of earth. Walls of structures should be protected either by frequent sprinkling or by hanging wet canvas over them. When temperatures are not extremely high, sprinkling of the concrete alone, if done at sufficiently frequent intervals, will often give the desired protection.

Mass work, such as foundation walls entirely below ground, heavy abutments, and retaining walls, do not require the same extreme measures of protection as does work of thinner section. Leaving forms in place and occasionally sprinkling or wetting down the work for several successive days will often be all that is necessary.

Roof slabs require essentially the same protection as thin concrete walls, since two surfaces are essentially exposed. Roofs should be covered with moist earth or sand, sprinkled freely, and otherwise be protected like pavements. Stucco work may be ruined by lack of proper protection while hardening. Much of the cracking and crazing of stucco has been due to neglect in protecting the work.

Joining New Concrete to Old

New and old concrete can be joined only with great difficulty and the strength of such a connection is always uncertain. It is only by using the greatest care that a cement finish coat can be made to adhere to a concrete base that has completely hardened. The joining of old and new concrete work is best done by thoroughly cleansing it from all dust and loose particles, sometimes chipping the surface, saturating it with water, painting it with a cement grout paint mixed to the consistency of thick cream, and while this coating is fresh, immediately placing the new concrete or applying the plaster coat as the case may be. As cement begins to harden within a very short time after being combined with water, the grout paint should be applied only a short distance in advance of the work going on. The more nearly clean aggregate faces are exposed to the new concrete or plaster to be applied, the better will be the bond secured between old and new work.

Concrete Surface Finish

Concrete surfaces are susceptible of a great variety of pleasing finishes that have practically no limit other than that imposed by individual ingenuity of the workers on the job. The types of surface finish may be obtained as follows:

(1) Leaving the concrete as it is when forms are removed;
(2) Using a mortar facing or plaster;
(3) Hammer dressing or tooling;
(4) Using special concrete mixtures;
(5) Coloring the surface;
(6) Washing the surface to expose aggregates selected with that end in view.

The kind of surface finish that will result from leaving the concrete as it is when forms are removed depends upon two factors. Such a surface will be truly characteristic of concrete if the forms have been well made of planed lumber and the concrete was carefully and thoroughly spaded next to form faces while being placed. If these essentials have been observed there is rarely or never any need of subsequent work on the surface for ordinary concrete structures. A few pebble pockets may be in the surface, and these can readily be pointed up immediately after forms are removed by using a mortar of cement mixed with the same number of parts of sand as were used in the concrete mixture. For example: If the concrete was a 1:2:4 mixture, then a 1:2 sand-cement mortar should be used for pointing. If a 1:3:6: mixture, then a 1:3 mortar should be used for pointing. This is necessary in order that the spots where the work is pointed up will not have a color different from the remainder of the surface.

A mortar facing or plaster may be considerably varied. Usually when such finish is intended for monolithic construction or for concrete block construction the concrete when placed in the forms is not thoroughly spaded next to form faces, thus intentionally leaving exposed pebble pockets on the surface which will insure a better key for the plasterer. In block construction where stucco is to follow the block are rough cast for the purpose of providing a better bond for the mortar coat. But the mortar coat itself is susceptible to considerable variation. It can be given different floated finishes by using a steel trowel, a wood float, or a wood float covered with a piece of carpet or burlap. Each of these methods of smoothing the finished surface produces a different texture.

The plaster coat may also be varied by tinting, but for such work only mineral pigments should be used because other colors lack permanence. In a plaster coat finish another variation consists of pebble-dash finish which is secured by throwing pebbles of uniform size, thoroughly washed and wet, against the soft mortar coat. To make these pebbles adhere better it is advisable to wet them with a thin cement grout immediately before throwing against the surface being treated.

Concrete surfaces can be tool or hammer dressed in just the same manner as stone surfaces are so treated. For successful hammer dressing or tooling, it is very necessary that the aggregates used shall be of uniform hardness throughout and have been selected in anticipation of such surface finish. If pebbles are used for aggregate and have considerable variation in hardness, the hammer dressing will cause some of the pebbles to break out of the surface and the finish will not be so pleasing as where aggregates of uniform hardness throughout are used. Also, the appearance of the work will be considerably influenced by the time which the concrete is allowed. to stand before being hammered. If too soft, particles of concrete will be loosened. If the surface is too hard, the work will take more time and hence be more expensive.

Surface finish of great attractiveness can be secured by using special concrete mixtures. These are usually facing mixtures in which the aggregates are selected sands and stone chips, the facing being placed at the same time as the mass of concrete by using a metal septum in the forms. After forms have been removed and the concrete has hardened, the aggregates may be exposed in a number of ways. The surface may be tooled if so desired, or it may be scrubbed down with water if forms can be removed within the first twelve hours or so before the cement is thoroughly hardened. If the work must be delayed until later, then an acid wash and scrubbing will remove the surface film of cement and expose the colored aggregate. The possible variations of colored texture to a surface, possible by using special concrete mixtures containing selected aggregates, is limited only by the variety of aggregates that may be obtained; also by the various mixtures that may be made by combining two or more of these aggregates.

White cement, white sand, marble dust, quartz screenings, mica spar, granite, marble chips of various colors, crushed, screened, graded selected chips singly or in combination are used in special concrete mixtures for various colored surfaces. In fact where colors are desired, this method of securing them is certain to result in permanence.

Aggregates are sometimes exposed by using a sand blast which removes the plastic or pasty effect given to the concrete by the forms, and produces a granulated finish somewhat similar to sandstone but not so uniform because the aggregates are likely to be irregularly exposed. For the hammering or tooling, pneumatic hammers are often used, especially where large areas must be treated in this way. Finishing with pneumatic hammers produces a very attractive surface for buildings in which concrete has appropriate architectural features. The bush hammer is used for intricate portions of the work and other plane portions are dressed with the pneumatic hammer. Owing to the cost of equipment necessary to produce sand blast finish, this method is not used except on expensive structures where the one item of finish does not bear a great relation to the total cost of the work.

In finishing concrete surfaces with carborundum stone, only the form marks can be removed by hand methods without excessive labor. Many irregularities cannot be entirely obliterated. A machine recently placed on the market is designed to remove all marks, giving a perfectly straight surface. The process is a dry one. The machine is essentially a revolving disc in which are mounted a number of hardened steel cutter wheels, which roll on the surface to be dressed, and remove the material by a chipping action. The disc is driven through a flexible shaft by an electric motor, carried by the operator. The cutting action can be made to give a surface resembling that obtained by a bush hammer, which offers a good bond for float or other finish.

There are also on the market several floor surfacing machines intended for use in finishing terrazzo floors, or in grinding down the usual cement floor surfaces where it is desired to obtain a finish similar to terrazzo.

CHAPTER 5.

FORMS FOR CONCRETE CONSTRUCTION

Care in Manufacture of Forms

No part of concrete construction requires more care than the making and use of the forms in which the concrete is placed to harden. The appearance as well as the safety of the finished work is governed in a large degree by the care used in making and erecting the forms. Frequently such work is so carelessly done that ends of upright supports or props rest on soft earth in such a manner that beams, girders or other portions of a structure sag out of true intended line. Often, sides of beam and girder forms, when these members are to be deep, are sometimes so poorly built and braced as to bulge or warp.

Forms supporting the concrete for floors frequently are not propped up with sufficiently strong timbers and so bulge under the weight of concrete, usually resulting in a continual sagging of the form while the concrete is hardening, thus resulting in permanent cracks on the under side of the slab. This naturally prevents the floor from having the strength for which it was designed.

Contractor responsible for forms

In all cases, care in the construction of the form work is very evident in the appearance of the finished work. Usually it will be found that the care and judgment displayed in building and erecting forms' will be proof of similar care exercised throughout the construction and the results of this, other things being equal, will insure a safe building of fine appearance.

Construction of forms should be left to the contractor since he has the responsibility for the work until it is completed. Naturally, however, there should be necessary cooperation between representative of the engineer and architect, so that no misinterpretation of specifications will be responsible for forms not in strict keeping with the work's requirements. The contractor's drafting forces should be impressed with the responsibility of careful design of forms, so that in every respect the finished structure may conform to the architect's and engineer's plans. In all instances the engineer should control the time for form removal on any engineering construction and his inspectors should be qualified to insure the safety of the structure by intelligent inspection of falsework and supports.

Many types of forms are used in concrete construction. Wood and steel are the materials most commonly used. Forms may vary from the most simple to the most elaborate, depending upon the nature of the work and its engineering or architectural details. Some forms are intentionally built for use once only. Others are and may properly be planned for repeated use, thus insuring considerable economy of form cost with respect to any one job.

Perhaps the most extensive use of steel forms is in connection with the manufacture of various classes of concrete products such as block, brick, tile, sewer pipe, concrete trim-stone, etc. Most of such forms or molds are in reality a part of the machine used to make or form the product. However, steel forms are used extensively in various other classes of concrete construction. Some of these forms are patented, perhaps with particular reference to a patent also covering a so-called system of construction. Many of them are, however, adaptable to any type of monolithic concrete work. Among the examples of use of steel forms are those used in connection with monolithic sewer construction, bridge arches, culverts and such circular structures as silos, tanks, chimneys, etc.

Some ornamental concrete products, such as garden furniture, trimstone and statuary, are cast in plaster, glue or sand molds. Such work, however, is generally confined to a central plant, therefore hardly comes in for description in what is intended to cover principles of field work of concrete form construction.

Form Economy

A great amount of material and labor is required in the construction of some forms on engineering structures. It is evident, therefore, that the cost of this work is not a small part of the cost of the finished building, and by exercising care in design much material and labor can be saved. It may be best in some cases to draw up sketches from which the carpenters who are to build forms can work and thus insure least waste of material.

The greatest economy is gained by building forms so that they can be repeatedly used. It is not uncommon to construct a ten-story building by using throughout the form work employed for the first three stories. This fact should be borne in mind in the design of reinforced concrete structures because it is frequently cheaper to keep the columns of the same dimension throughout in order to save the cost of changing forms. That is more true now than ever in view of the great increase in cost of labor and the scarcity of materials. Where beam and girder construction is used, consideration of form cost as against the use of a slight increased quantity of concrete will often determine whether it would be more desirable to reduce the size of section or leave it the same size as other corresponding sections of the building.

Economy in form construction results from devising ways and means to fasten sections together and in position on the work so that the least amount of nailing will be required. Every nail that is driven home gives trouble when the forms are taken down, and, because of careless handling, may result in much injury to forms, not to mention injury to concrete which has not thoroughly hardened, from the hammering and knocking necessary to dismantle forms. In many cases wedges, clamps, ties, brackets and a variety of fastening devices other than nails can be used, thus considerably reducing the labor to set and take down forms and also resulting in prolonging their life for use a greater number of times.

Safety Dependent on Form Construction

The safety of a reinforced concrete floor, as well as that of the entire structure, may be jeopardized by faulty form construction. Many of the failures of reinforced concrete, if not most of them, can be traced directly to weakness in the forms used, due to faulty construction of their supports or the false work. For example, if forms for beams and girders are supported by struts of insufficient strength, causing a collapse of the form in one part, the probability is that the forms of the whole floor will collapse with it and deposit its contents on the floor below. A load suddenly applied in this manner to the floor below might cause collapse of the whole building, since the accident would probably take place at a time when none of the concrete had attained the strength ultimately to be realized.

In order that there may be no doubt regarding the safety of forms, they should be carefully inspected before any of the concrete is placed. Upright supports and braces must be examined to determine that they are sufficiently strong to carry the weight of the wet concrete. Frequently, false work and braces are improperly placed and do not have sufficient support. It is also not uncommon to find forms braced against green masonry and brick work and struts secured with a few wire nails where they should have been notched into supporting timbers and well spiked in position.

Standardization of Panels

In the construction of forms, it is often possible to standardize a set of panels that will permit of considerable latitude in setting up, thus making re-use of the same set of forms possible a large number of times. Sometimes such panels are metal lined with galvanized sheeting in order to prolong their life and at the same time insure a better appearing finished surface to the work.

Wood Forms

Where wood forms are used most of the work of cutting and erecting forms is done on the building site. However, for some classes of work such as foundations for relatively small structures like dwellings, standard panels are used and if carefully handled may be re-used a number of times. On account of the cost of form construction, any successful attempt to standardize sections so that wood forms can be used repeatedly is likely to prove profitable. Some of the various so-called form systems which have already been mentioned are adaptable to a variety of construction uses other than the particular system of construction for which originally devised.

The kind of lumber to use for form construction depends to a considerable extent on the kind available and whether the material after first use is to be again used in similar work; also, whether the building is to have many floors of the same design and construction or whether only two or three stories. Ordinarily, good material should be used in form work, for it permits good carpenter work and also smooth, true concrete work. To support forms it is the custom to use as cheap a grade of rough, sound lumber as can be purchased. For the sheathing, or that portion of the form in contact with the concrete, it is best to use a good grade of well seasoned hard pine because such material possesses both strength and grain of sufficient closeness to prevent the form boards from splintering badly. Norway pine, when obtainable, is one of the best materials. The hardwoods are too costly and too expensive to work. The softer woods do not permit of repeated use, although for exceptionally fine surface finish and detail, it is sometimes necessary to use soft grades of pine because of the ease with which carpentry can be done on the forms to reproduce the surface wanted.

If the forms are to be used once only, sheathing is generally made up of 1-inch dressed material. If they are to be used several times, the side and bottom form boards of beams and girders are made of either 1½ or 2-inch dressed planks. Shores and supports for the centering generally consist of rough 3 by 4-inch studding. Timbers as large as

4 by 6, 6 by 8, and 8 by 10 may be used to brace and secure form work where the masses of concrete to be supported are exceptionally heavy.

Sliding Forms

In some instances sliding forms are used in certain classes of construction, notably in the erection of circular grain tanks or other buildings which it is intended shall be truly monolithic throughout. Such forms, as the name implies, are kept moving continually upward as concrete is placed. In general, sliding forms consist of wide outer and inner panels, operated by means of jacks which derive their support from vertical rods or pipes embedded in the concrete. The use of sliding forms requires continuous placing of concrete for twenty-four hours daily so that the forms may be moved upward slowly at a steady and uniform rate, thus preventing concrete from adhering to them. In the use of silding forms it is essential that at all points the forms be kept in exact horizontal and vertical relation to each other. These requirements indicate that although the system is quite simple, experience and constant watchfulness must be on the job to obtain the best results. Also the system of jacking up the forms must permit of positive control and there must be no slipping of the jack on its support.

Sliding forms are usually made fast to an upper and lower wale, consisting of 2-inch planks with overlapping joints. In circular concrete construction, one edge of these planks is cut to a radius of the wall surface plus or minus the thickness of the sheathing. Therefore, there is an inside and outside pair of wales cut to segments corresponding to the curve of the outside and the inside of the wall respectively. To these built up wales, 1-inch sheathing about four feet long is nailed usually with triangular bracing between the wales to give the forms required rigidity. The yokes are attached at intervals to these forms and may be entirely of wood, although the head piece is often made of steel plates so as to insure little variation in the separation of forms and to give a rigid yoke member to take the lifting action of the jacks.

Jacks used are of several different types, some of which are patented. Two types are common. Both of these work on the jack screw principle. A common type of jack consists of a hollow screw fitted at its top with a turning head and at its bottom with a device commonly known as a ratchet, or "dog," for gripping a vertical jack rod which is embedded in the concrete. The jack rod passes up through the center of the jack, which in turn passes through the threaded head piece of the yoke. By turning the turning head with a bar, the yoke is caused to move up, taking the forms with it. Successive turnings of the jacks through the same angle will thus cause every portion of the forms to move the same distance. A strict regulation of the amount of turning which each turning head is given is absolutely necessary to correct operation of the forms.

Another type of jack consists of a threaded rod fitted with a collar which bears on a 1¼-inch pipe embedded in the concrete. The threaded rod passes through the head piece of the yoke and is fitted with a turning head. The pipe is made in 3-foot lengths, cut square at the ends and without threads. The lower end of the threaded jack rod extends down through the 3-foot length of pipe and 6 inches into the pipe below. This keeps all of the 3-foot lengths of pipe in line, while the collar bearing on the top of the uppermost pipe furnishes a positive support to the jack. When the forms have been raised the full distance of their travel for one position of the jacks, the clutch is released, the jack moved up and a short length of rod or pipe, as the case may be, placed on top of and in line with those already embedded in the wall. The ratchet then engages the rod and the work proceeds. When rods are used, the successive short lengths are connected by sleeves.

Sliding forms have been used for a considerable variety of structures. These include standpipes, water tanks, grain tanks, grain elevators, mill houses and chimneys. The system is also adapted to the construction of hollow concrete piers and warehouses or walls, although it has so far been but little used in such construction. When such forms are used by experienced contractors in the construction of mill and elevator buildings, window openings can easily be provided for and cross beams from one wall to another also can be constructed without special difficulty.

Dry Form Lumber Desirable

Form lumber should be free from shakes, rot and knots. Knots and similar imperfections leave their imprint on the finished surface of the concrete and in addition weaken the lumber, hence the forms. Air seasoned lumber is better than kiln dried. The latter will swell and bulge at the joints, while green lumber will shrink if not kept wet, resulting in the opening of cracks through which water carrying cement will leak when the concrete is placed. Even for rough work, lumber that is dressed at least on one side and two edges is best for form sheathing so as to make the boards fit closely together. The planed surface will also reduce the labor of removing and cleaning forms. Tongued and grooved stock, as well as ship-lap is often used for sheathing. Beveled edge stock is preferred by some because if the lumber swells, the edges will slip past each other without causing a warping or bulging of the boards. Form sheathing should be of uniform thickness to prevent unevenness in the concrete surface.

Cost of Forms

The cost of forms varies within a wide range. One of the most accurate methods of estimating the cost of wood forms is to figure the board measure of lumber and the hardware required, the carpenter labor per thousand board measure to erect, the labor of taking down and cleaning forms, and the cost of transporting lumber to and from the job. A certain percentage of the cost of form lumber should be charged to the job. What this shall be can be learned only from carefully compiled data covering a number and variety of jobs that will disclose how long forms may be used repeatedly. The amount of skilled and common labor required to make, erect and dismantle forms, depends upon how complicated they are and the conditions under which the men work. The amount of carpenter labor per thousand board measure of lumber for form work is often difficult to estimate. A carpenter experienced on concrete form work can accomplish more than one whose experience has been confined to general building construction. Experienced form carpenters bear in mind that forms must, if possible, be designed and built to permit salvage and further use, therefore, do no more cutting of lumber than necessary. They also plan the forms so that they may easily be taken down with least damage.

Along with all other commodities lumber has increased in cost. Consequently, it costs more to make wood forms than it used to. For that very reason greater care in making forms with respect to cutting and fitting the various pieces as well as greater care in erecting and dismantling forms is likely to pay better than ever. It should be possible to use forms several times and the greater the number of times they can be used, the lower the percentage of original cost to be charged against any one job. For certain classes of construction, form lumber can be used ten times. Seldom should more than twenty-five per cent of the first cost be charged to one use unless there is an unusual amount of cutting of stock lengths.

Wetting Forms

In dry, hot weather it is common to wet down forms immediately before placing concrete in them. This assists to prevent concrete from sticking to the forms and also keeps the forms from absorbing water from the concrete, necessary to its hardening. Each time after taking down and before using again, forms should be thoroughly cleaned of all adhering concrete.

Centering and Falsework

The construction of centering and false work for reinforced concrete bridges must be sufficiently strong to carry the weight of the wet concrete of the completed structure without deflection or movement. All such form work must be securely braced against failure from the pressure of concrete. The principal requirement, however, is that the centering be so arranged that it may be readily dropped away when the concrete has hardened sufficiently to be self-sustaining and it is desired to remove the forms. Probably the best method is to use wedges of hardwood under main uprights or under the false work supporting the lagging, then these can be backed out, thus allowing the centering to settle under them. Another method is to use what is known as a sand box, which consists of a tight metal box fitted with a plunger upon which the uprights stand. The box is filled with sand and in it is a small aperture which can be closed with a screw plug or some other way. When desired to drop the centering, the aperture is opened, allowing the dry sand to flow out, thus permitting the plunger and centering to settle. Sometimes centering is built on screw jacks and where several arch rings are to be built parallel with each other, the entire centering is built on trucks or rollers carried on a track. In this way when one arch ring has been completed, the false work may readily be shifted to the new position.

CHAPTER 6.

USE OF REINFORCING STEEL IN CONCRETE

Need for Reinforcement

Reinforced concrete is defined as a combination of concrete and metal, preferably steel of certain determined quality, the metal being so placed as regards position and quantity that the concrete and the metal both take and resist the strains which they can best withstand.

Concrete is about ten times as strong in compression as in tension. Plain concrete—that is, concrete without any reinforcement in it—would have to be unnecessarily massive to possess the required strength as subjected to tension in construction. As a matter of fact, it would be found impossible to design most structures of concrete were it not that steel is embedded in it to provide the tensile strength which concrete lacks and at the same time fully utilize its compressive strength. In mass work where the load is placed immediately upon concrete—that is, where the load is one of compression entirely—reinforcement is seldom if ever used, but great economy of design as well as the great utility of concrete has resulted from proper disposition of reinforcement in it.

Quality of Reinforcing Steel

Any quality of steel will not do for reinforcement. Usually a certain quality is specified, that is steel having a certain chemical composition and given physical properties. Such steel, if properly embedded in concrete of correct proportions and consistency so that everywhere the concrete will bond or adhere perfectly to the metal, takes the pulling and bending strains because the adhesion between the concrete and the steel prevents the latter from slipping in the mass. These facts make it possible to use concrete in many ways that would be impracticable were it not reinforced.

Position of Steel

In the design of concrete structures, the quantity and position of steel are indicated on the plans. It is important that no less a quantity be used than called for and still more important that the position of the steel as shown be strictly secured in the process of construction.

The position of steel is determined by the nature of the loads to which the structural member or portion of the building is to be subjected and also to insure that it is covered with sufficient concrete to protect it from prolonged exposure to fire and from corrosion due to possible penetration of moisture.

Piers

In piers, posts and columns the concrete takes compression assisted by the steel, and the vertical steel takes tension if any bending occurs.

Beams

In beams there are stresses of compression, tension and shear. The concrete takes all the compression and a limited amount of shear. The steel is computed as taking all the direct tension and assists the concrete to carry shear stresses.

Columns

It is customary to reinforce all columns whether reinforcement is theoretically required or not. The practice is to use at least four rods so disposed in the column that their location corresponds to the four corners of a square. In no case should the steel be nearer the surface than one and one-half to two inches, since it is desirable to prevent any tendency of the vertical rods to buckle under eccentric stresses or working loads. It is also customary to place occasional horizontal hoops or ties around the vertical steel. Such hoops in the form of wire ties are helpful in holding the rods in correct position while concrete is being placed. In some systems of reinforcing columns, reinforcement is shipped to the job practically assembled. The same is true in part of reinforcement for certain types of beams or girders.

Types of Reinforcements

In addition to plain round, plain square and twisted square bars, there are various types of deformed bars used for concrete reinforcement. In general, the deformation consists of lugs or other projections formed on their surface during the process of rolling. The principal object of such deformation is to increase mechanical bond and to safeguard against the effect of absence of mechanical bond in places where concrete may not everywhere surround or be in contact with the steel.

In addition to steel bars of various forms, there are many so-called metal fabrics or meshes used as concrete reinforcement. The material commonly known as "expanded metal" for example, is formed by slotting sheets of steel having required thickness and physical properties, then stretching or expanding the sheet, thus opening the slots and considerably increasing the area of the sheet. Some types of expanded metal are particularly adapted to exterior and interior plaster work such as stucco. Some others are intended to be used so that they will make actual forms unnecessary, even to secure monolithic construction. For example, a steel frame is set up, expanded metal or so-called metal lath attached to each side of this frame, the space between lath filled with concrete and finally both outside surfaces plastered, thus giving a monolithic reinforced wall.

Other types of mesh reinforcement may be likened to woven wire fencing or to square mesh fabric similar to that used for common screens. These vary in weight per certain area due to the weight of strands of wire used in forming the fabric. Some such fabrics are woven, others are welded where the strands intersect or cross.

It makes little difference which one of these forms of reinforcement is used in certain classes of work, other than that perhaps in specific cases some one type will be more economical or may be otherwise used more advantageously. The principal thing to observe is that the physical properties of the steel and the net cross-sectional area of steel are satisfactory and the material is in the proper place to take all the tension to which it will be subjected.

Handling Reinforcement on the Work

Often on the job, reinforcing metal is carelessly thrown about and allowed to become rusty and perhaps covered with oil or other filth. Care should be taken when placing it to make certain that it is free from mill scale or scale in the form of rust. Either of these can be removed by pickling in weak acid or perhaps by brushing with stiff wire brushes.

Frequently some steel must be shaped on the job to conform to the details of design shown on the building plans. Various devices are used for bending steel, typical among which are Koehring bar benders. These are made in two sizes. The smaller will bend bars cold up to one inch square. The bending die is two inches in diameter, giving a one-inch radius to the inside curve of the bar at the bend, and has roller-bearing journal. The guide block is provided with a large roller, allowing the bar to follow freely, in order to avoid fracturing by too great tension. This roller reduces friction to a minimum.

The large type will bend up to 1½-inch round or 1¼-inch square. The bending die is 3 inches in diameter, which gives a 1½-inch radius to the inside curve of the bar at the bend, and likewise has roller-bearing journal. This bender is provided with an adjustable clamp, which automatically adjusts itself to the thickness of the bar, and securely holds the bar at one end. It is so constructed that the small bars can be bent with direct leverage, and bars one inch and larger can be bent with two men by using the back gear attachment. The pinion is provided with a ratchet lever, and this pinion engages the gear segment on the main frame.

The bending point is central with the journal, which gives it greatest leverage, and this combined with the roller bearing journal and roller guide makes it possible to bend the bars with least force. Any size bar up to 1½-inch round can be bent without making any adjustments. The bender also affords advantages of counterbending, without removing the bar or without making adjustments in case a bar has been bent too much.

It is also necessary to have a device capable of cutting bars as may be required. One of the most efficient tools for this purpose is the Koehring bar cutter, also made in two sizes.

In designing this bar cutter the primary object was to provide a machine light in weight, convenient to handle, effective in operation, and simple in construction, so it will not get out of order.

The cutting jaws are so arranged that it becomes more powerful as the resistance increases in advancing the cut, and so that the bar to be cut will not be pushed away by the two knives in the cutting action. When cutting, the two knives advance, which makes a clean, square cut, and does not leave fractured ends.

The cutters are built entirely of steel, and include handle for operating. With the No. 1 size one man can easily cut a ¾-in. square twisted or round, and with the No. 2-A size two men can easily cut a 1¼-inch round or a 1⅛-inch square twisted bar.

Stops are provided to prevent the bar from twisting, and in cutting the various thicknesses, care must be taken that the bar is placed at right angle with the knives and that a spacing block is placed between the bar and the stop post. If the space between the bar and the post is not filled in, it will allow the bar to twist while the knives are in cutting action, and this will cause them to break out. See that the bar is placed straight through the opening at right angle with the blades, and that a piece of wood or iron is placed between the stop post and the bar.

In shaping steel by bending, care should be

taken not to exert the bending strain suddenly or violently, but rather gradually and uniformly to prevent any fracture at the point of bend.

Makeshift Reinforcement Dangerous

No doubt concrete failures have resulted because of makeshift practices in reinforcing. Old chain, wire cable and similar scrap material, which seldom can be handled to advantage in the forms, is quite likely to be ineffective even though the sum total of metal used is greatly in excess of actual requirements determined by design. Generally this is due to the fact that once a strain of tension comes on the concrete greater than it can bear the immediate help of reinforcement is lacking because the steel or other metal was so placed as not to take tension immediately.

It is necessary that the action of reinforcement be continuous, and as in some classes of work bars cannot be secured of sufficient length to extend between two desired points, it is necessary to splice them. Splicing is usually done by lapping the bars a certain number of times their diameter, roughly 50 to 60 times this diameter, at a point of least stress. It is common where laps are necessary in rectangular structures, to make them at or near the center of one side rather than at or near a corner.

Examples

Two simple examples will serve to illustrate the results obtained by hoop reinforcing of a column. Make a cylinder of thin paper and fill it with sand. The paper may be strong enough to hold the sand if not disturbed, but if a load is put on top the paper will burst and the sand will flow. If the shell of a tin can, when top and bottom is removed, is used in place of the paper cylinder, it will take considerable load to burst the tin confining the sand. If, instead of sand, the cylinder is filled with cement mortar or concrete, it can be seen that the concrete will take its full measure of load in compression and will have the benefit of the encircling tin to prevent crushing or failure when the safe compressive load of the concrete has been approached or exceeded.

The principle of reinforcement in a beam can be illustrated in a very simple manner. If a column of concrete is made 6 by 12 inches in square dimension and 10 feet long, then laid in such position that its two ends only are supported, it corresponds to a beam. Realizing that concrete is relatively weak in tension, it can readily be seen that it would easily be possible to make this concrete beam break of its own weight by increasing its length sufficiently between supports. The neutral axis in a beam is the point where the character of stress changes from tension to compression or from compression to tension. In a beam of homogeneous material, that is, one in which the tensile and compressive strengths are equal, with symmetrical cross section, the neutral axis will be midway between the top and bottom surface, or skin. At this point stress is zero. At other points throughout the section it will vary in amount and nature (tension or compression) with reference to the location of the neutral axis.

It is not the purpose of this discussion to elaborate on the subject of reinforcing concrete since it is a very technical one and requires a thorough knowledge of engineering for a full understanding.

Frequently reinforcement is used in some parts of structures not because an increase of structural strength is necessary, but to prevent unsightly cracks due to volume changes in the concrete from changes in temperature, in other words, from expansion and contraction. Such reinforcement is known as "temperature reinforcement."

The Koehring 283
Heavy Duty Construction Mixer
mounted on skids.

CHAPTER 7.

NOTES ON SPECIFICATIONS

Specifications Should be Clear

Specifications for any construction work of importance are necessary so that there may be a clear understanding on the part of the contractor as to the methods by which the work is to be carried on to conform with the engineer's plan.

The first requirement of any set of specifications is that it shall be in such form as to present only legal demands and afford legal protection to all persons interested. This means that a specification should be so worded as to be easy of interpretation. It should not contain phrases or expressions that by any chance might be given more than one interpretation.

Technical terms and phrases must necessarily be a large part of the text of specifications. Such terms are well understood by the engineering and contracting professions, yet it is possible to so refer to technical subjects that a clear understanding of them may be had by anyone who can read. It is well in the case of some unusual terms or expressions to append to the specifications, if necessary, a glossary of terms that will define such words or phrases in the light in which they are to be interpreted in reading the specifications.

Responsibility should be Clearly Defined

Specifications should not be drawn in such a way as to release the engineer from responsibility justly his. If for any reason there is wisdom in dividing responsibility, the specifications should be so clearly worded as to indicate what responsibilities rest solely on the engineer or on the contractor, and what responsibility is to be shared between them.

Another important requirement of specifications is that they shall demand only methods that are generally recognized as practicable—that is, no unreasonable demands or requirements should be made. Also, it is poor practice to write rigid specifications if there is no intention of carrying them out literally as written.

There probably never will be a piece of work where minor adjustments will not be necessary from time to time as the work progresses, but this does not mean that the specifications applying are not practical. Specifications should be so defined that the contractor may know as nearly as possible the grade or grades of materials he will be expected to furnish, the quantity of work he will have to do and as much other accurate, necessary information as possible. They should be so drawn that neither a superficial analysis of them, nor an attempt to carry them out, will prove that they are at variance with the plans on which they are based.

Features for Consideration

Among the various features of work that must be given consideration in drawing specifications for plain and reinforced concrete construction, are the following:

Cement
Aggregates
Mixing water
Reinforcing steel
Proportioning, mixing, and placing concrete
Type of mixer to be used
Requirements for placing reinforcing steel
Methods of bending steel
Work done in cold weather
Form construction
Protection of concrete, both in cold and warm
    weather
Removal of forms
Patching or repairing imperfect concrete
Provision for expansion joints

Inspection

Status of various men on the work as between representatives of the engineer, contractor, owner, etc.

It should be remembered that many large pieces of work often involve changes of one kind or another in the original plans. There should be a clause in the specifications that will state how such changes are to be handled by the contractor, particularly with regard to additional or decreased compensation.

Often it has been the practice to arrange for the arbitration of disputes in a way that was not always equitable to either or any of the parties interested—that is, the arbitration board provided for in the specifications frequently included at least one member likely to be partial to one or the other of the parties most concerned.

From the very nature of concrete construction, whether plain or reinforced, it is difficult to make alterations in the work after it has progressed beyond a certain point. Therefore it is important that provisions be made for continuous inspection as the work advances, so that before any considerable amount of it at variance with specifications can be finished it will be possible to correct faults.

Basic Principles in all Specifications

As nearly every piece of work has some features peculiar to itself, specifications vary in accordance with the particular job for which written. However, certain basic principles of concrete work are a feature of every job. Therefore certain portions of every specification are alike for all jobs. The extensions of, or additions to, a specification may be almost without limit, yet should not be made without some definite purpose in view.

Nearly all specifications that form the basis of bids for certain pieces of work are accompanied by explanatory notes, consisting of instructions and information for bidders. These items usually prescribe the method of making a bid and how it should be forwarded when ready for delivery to the person or persons interested. If the bid is accepted, the next step is the signing of a proper contract which in its essentials states when the contractor will commence the work and the time he will be allowed for its completion. All specifications for work likely to be subject to various delays beyond reasonable control, should contain a clause outlining the procedure necessary to obtain extension of time on a contract.

On any job the first work usually involves more or less excavation or other preparatory work. On large jobs, contracts are usually let for various parts of the work so the concrete contractor may have nothing to do with excavation or other preliminaries. If so, the contract should so state.

The first paragraph of a specification for concrete work involves the cement to be used. As a rule, most specifications are too wordy in laying down the requirements for the Portland cement to be used on the job—that is, they go into such details as stating the specific gravity, fineness, time of setting, tensile strength, chemical content, etc. These requirements can be met by briefly stating that the Portland cement used shall meet the requirements of the latest Specifications and Tests for Portland Cement of the American Society for Testing Materials. If this is done, any of the well-known brands of Portland cement, the manufacturers of which have subscribed to these specifications, may be relied upon to meet the demands of any concrete construction.

Another objection to specifying in detail the qualities that the cement shall possess, is that very few persons, other than those whose experience in sampling and testing cement has been acquired by thorough laboratory work, are competent to undertake cement testing. This fact has often been responsible for high grade cement being rejected on the job. If for any reason it is desirable to test cement, the work should be intrusted to a well-known, well equipped testing laboratory which specializes in such work.

Aggregates

The second item of the specification usually covers the subject of aggregates. These will consist of sand and pebbles or of sand and some kind of crushed stone, among which may be slag or other materials of rock origin. If fine aggregate such as stone screenings is to be used in place of sand, or if cinders are to be used in place of other coarse aggregate, the specifications should detail very carefully the quality of such materials and should give field methods of determining their quality. Actual tests on aggregates, like tests for cement, should be made in some laboratory properly equipped for such work.

The specifications for sand, usually referred to as fine aggregate, are becoming more and more rigid, and rightly so, since it has been proved that the quality of the sand affects the quality of the resulting concrete in a far greater degree than was once realized.

Proportioning Mixtures

The next item of specification should take up the subject of proportioning mixtures. It should go explicitly into the manner in which the several ingredients are to be measured and mixed, so that by no chance can the specification be interpreted to mean that a 1:2:4 mixture and a 1:6 mixture are identical. Methods of measuring materials should be definitely stated, as should the methods of mixing them.

Type of Mixer

Engineers are in accord on the point that only a batch mixer should be used. Specifications should be specific in stating how and in what type of machine materials are to be mixed. The time of mixing, or its equivalent in number of revolutions for particular make of machine, should be carefully specified and rigidly enforced.

Definite amount of water should be used for each successive batch so long as materials are constant in physical properties and moisture content.

The importance of the correct amount of mixing water is another thing which has but recently been thoroughly appreciated. A clause should describe the consistency so clearly that any excess of water will be avoided.

In reinforced concrete work, methods of placing reinforcing steel and all practice involved in handling and bending it where necessary to form it to shapes shown on the plans, should be clearly stated in the specifications.

On many jobs it is necessary to devote considerable space in the specifications to detailing methods of making and erecting forms. A great deal of concrete work has failed to attain the strength that might have been expected had proper attention been given to form construction and all details concerning erection, use and removal of forms.

Concrete work is often carried on during low temperatures. In fact, during recent years, many large contracting firms have practically specialized on winter work. They have proved conclusively that good concrete work can be done in cold weather. But in anticipation of a certain contract extending through a season when temperatures will be near or below freezing, specifications should clearly define how work should be handled under such conditions.

Much of the success of finished concrete work depends upon carefully protecting the concrete for a certain time after placed. This is true not only in cold weather but also in extremely warm weather, especially when hot, dry winds prevail. This latter fact is seldom appreciated and the practice of protecting concrete in warm weather is almost universally neglected except in highway pavement construction.

There should be a clause in every specification covering such protection as it is desired shall be ˙ven the work for a stipulated period under certain conditions. For certain classes of work, the time when forms may be removed should be definitely specified.

If surface finish other than that secured by placing the concrete in the forms is desired and the same contractor is to attend to this part of the work, the specifications should state the kind of surface finish and how it shall be secured.

Sometimes concrete work is carried on under specifications which require that proprietary preparations such as waterproofing mediums or floor hardeners be used in some parts of the work. It is not right to place entire responsibility on the contractor for the success or failure of the work as a whole, where the use of such mediums is made a part of the specification. In these cases responsibility should rest upon the manufacturer of the particular preparation and he should supply and enforce the specifications governing, thus fixing responsibility for success or failure.

No specification should be longer than necessary to cover all details of the work, yet far too many specifications fall back on such expressions as "Shall be done in a thoroughly workmanlike manner" or "Shall be done to the satisfaction of the engineer." These expressions are vague in defining quality of workmanship and are sure to lead to almost interminable disputes on some occasions.

Considerable trouble on every concrete job of any magnitude, where the job naturally comes under the jurisdiction of local building ordinances, is due to the fact that plans as well as specifications frequently are at variance with building code requirements. Such trouble can always be avoided if specifications and code requirements are harmonized before work is started.

An interesting picture, showing complete plan of roadbuilding operations, from the material bins to the finished concrete road.

CHAPTER 8.

ESTIMATING COST OF CONCRETE CONSTRUCTION

Cost of concrete construction depends upon many conditions. Cost of labor and materials varies widely with locality.

In making estimates upon reinforced concrete structures, cost may be closely approximated by taking a unit price per cubic yard or per square foot of floor surface from similar work recently completed under nearly or practically identical conditions. Such unit price may be used for approximating the cost of a proposed structure or piece of work. In order, however, that such an estimate will approximate probable cost, the plans and specifications as well as prevailing prices of labor and material must be carefully compared with those of the structure or job from which the unit prices are derived. Even then it must be realized that such an estimate can only be an approximation. It may, however, be a very close one if the contractor has had considerable experience and throughout his experience has made it a practice to carefully compile and analyze cost data for the purpose of enabling close comparisons. Cost figures, based on one's own experience, are much more valuable than those derived from the experiences of others, as the latter relate to unknown conditions of working, while the figures which the individual compiles from his own experience relate to his way of working and the conditions which he has met and with which he is therefore most familiar.

Estimating the cost of concrete construction is in many respects different from estimating cost of other classes of work. Naturally, experience in handling a particular kind of job qualifies one to estimate more closely on that kind of work than on another kind with which one has had little or no experience. Frequently the inexperienced contractor has found out that the knowledge which he thought he had acquired under the direction or guidance of someone else, has not been so substantial as he believed, in that he lacked the inside viewpoint and, therefore, was unable to make careful analysis of all factors affecting cost.

Items Entering into Cost

The following items enter into most work and are therefore subject to detailed consideration before the preliminary cost of a piece of work can be considered as complete:

Interest
Overhead expense
Bonds
Insurance
Employer's liability
Public liability
Fire insurance
Special risks
Freight on outfit to and from work
Moving in and out
Land damages
Camp and other buildings
Preliminary work
Cost of materials plus freight
Hauling materials
Supervision
Labor
Loss due to camp operation
Transportation of labor
Tools
Plant rental
Running repairs
General repairs
Specials
Machinery
Fuel and power
Weather

The subject of interest includes proper return on the actual investment in equipment, as well as money that must be paid to any source from which financing for a job must be drawn. Certain jobs can be financed with a sum not exceeding 10 per cent of their total cost, providing the contractor succeeds in making suitable arrangements for payment on account, as the work progresses. Financial standing and credit rating of a contractor regulate his activities, and limitation of these frequently causes failure.

Overhead

Overhead expense is made up of a number of items and should include salary allowance for members of the organization, traveling expenses and similar incidentals. Under overhead would also come the maintenance of an equipment storage yard.

Bonds are necessary to safeguard the community or individual owner with respect to faithful performance of the work in accordance with the specification and contract.

Insurance

Insurance that must be provided is of various kinds and includes protection from fire loss on any combustible materials, safeguards against damage to nearby buildings or structures, protection against damage claims due to bodily injuries of employes and any other special risks against which the contractor may be protected by underwriting.

Many contractors underestimate the amount of freight or other transportation charges involved in moving their outfits from one job to another.

Cost and Quantity of Materials Very Important

In making up an exact estimate, the cost of the quantity of materials required can be accurately determined by using the figures corresponding to current market prices. The materials used in concrete construction may represent anywhere from 20 to 70 per cent of the total cost of the work. It can readily be seen that in so large an item, errors in estimating may affect profit considerably. The contractor must also know that the materials delivered to him are up to quality and quantity required so that he will not be called upon to remove or demolish a portion of completed work or secure additional materials beyond those delivered in insufficient supply.

In the case of practically all other building materials, finished products of known quality are supplied to the contractor and he is simply a builder; but in concrete work he is a manufacturer also, and must be competent to correctly judge the quality and suitability of materials entering into his work.

Labor costs can be estimated accurately only when experience has proved the volume of work which certain kinds of workmen or laborers may be depended upon to perform in a given time or under given conditions. Labor costs in erecting forms are particularly difficult to determine, especially where reinforced concrete construction is new to a locality and the only carpenters obtainable have had little or no experience with such class of work.

Speed of construction influences cost and in turn is influenced largely by the availability of space and labor. Frequently throughout a job a contractor is confronted by a shortage of labor or unsatisfactory labor. Competition of nearby work may make it difficult for a contractor to maintain or retain an efficient organization.

The uncertainties of transportation and its cost affect the cost of work within a wide range. It must, if possible, be arranged that there shall be an uninterrupted supply of materials in order that the work may progress without costly stoppages.

Weather Conditions

Weather conditions play an important part on the cost of work. It is not possible to foretell when storms or unfavorable weather may interrupt the work, except that experience shows that more dependable weather prevails at certain seasons of the year than during other times. Late fall and early spring are marked by unsettled weather, and if the work must begin in or extend into either of these seasons, calculations must be made on the interruptions that will result and the probable influence in cost that they will have on the work.

A comparison of different classes of work will show which factors most influence cost as among various classes. In reinforced concrete construction, for example, cost of materials is secondary to that of labor, while in plain mass concrete construction there is a larger cost for materials than for labor. Some classes of work call for complicated and therefore expensive form construction. On other jobs forms represent but a relatively small portion of the total cost.

Relation of Speed to Cost

Speed of the work is a factor in estimating cost. The work may be carried on too rapidly at certain points, thus causing congestion of labor and materials. For the same reason cost will be affected favorably or unfavorably by a well or poorly balanced construction gang. The mixer and associated equipment may be too small or too large for the number of men working or there may be too few or too many men for the equipment.

Often contractors must accept notes, bonds or other commercial paper instead of cash in payment for their work. The market value of such paper should be definitely determined. When possible to do so, arrangement should be made to dispose of securities of this kind at a definite price before closing the contract. Only by knowing what such securities will bring in cash can the contractor safeguard himself in estimating. For example, if he is given negotiable securities having a face value of $10,000, which will bring only $9,000 cash, his estimate must include an amount sufficient to cover such difference between face and market value of securities accepted as cash.

Cement requirements can be estimated accurately. It is, therefore, not necessary to increase the quantity estimated other than to add a small sum to actual cost per barrel to cover sack losses. The percentage of sacks lost is largely dependent upon the care received while in the hands of the user. On many large jobs it is possible to adapt bulk cement to the work and avoid the temporary investment and the certain loss of some of this investment when cement is used in sacks. It is always best to have one or more men, as necessary, made responsible for the careful handling and bundling of sacks.

Aggregates are sometimes sold by weight and sometimes by the cubic yard, or a unit weight is adopted and deliveries in cubic yards are converted into equivalent weights. If aggregates or cement must be transported and rehandled from trucks to cars and cars to trucks, where there is likely to be loss, there is also the cost of this transportation and rehandling. There is also some wastage of materials, particularly aggregates at the site of the work. It is, therefore, safe to estimate 10 per cent over actual figured aggregate requirements to cover such loss.

Cost of Water Should Not Be Overlooked

The probable cost of water is often overlooked. Frequently there is no difficulty in arranging for a sufficient supply. Within city limits, for example, the necessary arrangements can be made with the city water department for hydrant or similar connection. The quantity of water required may safely be figured at from 40 to 50 gallons per cubic yard for concrete only. In addition to that there is the water required for operating mixers, engines and other equipment, for wetting down forms and sprinkling concrete while hardening and for incidental waste that may bring the estimated requirements up to 100 gallons per cubic yard of concrete in place. The problem of water supply is often a troublesome one on highway construction. Frequently the supply must be hauled in tank wagons or piped for long distances, and before estimating cost of water, the situation to be met must be carefully analyzed. Hauling charges involve loading and unloading of materials, including working as well as waiting time of team and driver or motor truck and actual time of truck in travel.

CHAPTER 9.

NOTES ON CONCRETE CULVERT AND BRIDGE CONSTRUCTION

Culverts

The simplest form of concrete culvert is that made of precast pipe. It is adapted to all sizes of opening from 12 inches upward to the largest size of pipe made, providing the largest size will otherwise suit the situation. Good practice limits the minimum size of waterway openings to 12 inches because smaller sizes easily become choked with rubbish.

The box culvert is the most generally used of all concrete culverts because for the head-room the greatest area of waterway can be secured, and very simple forms are required. As the name implies, it is merely a long box with concrete top, sides and bottom. It is in effect a small concrete bridge with top slab acting as a floor to support the loads of traffic. The slab must be reinforced with steel rods or heavy mesh fabric. In building box culverts, a concrete floor should be laid in the bottom of the culvert. This is sometimes omitted and the sides extended down a short distance into the stream bed. It is always best to put the floor in to prevent damage from undermining.

The arch culvert is different from the box culvert in that the top is in the form of an arch instead of a flat slab. There is no advantage in the arch culvert for small spans. It requires little or no reinforcing because the concrete takes all of the load in compression. Form work for arch culverts is more costly than for box culverts.

The area of waterway for culverts is shown in the following table. These figures are given merely as a guide from which an estimate can be made of the approximate size of opening required. The area of the waterway opening depends upon the watershed to be drained and the amount of rainfall which must be taken care of from this area.

Table No. 13

SIZE OF WATERWAY REQUIRED FOR
VARIOUS AREAS TO BE DRAINED
(From Bureau of Public Roads)

Area Drained	Area of Waterway Needed (in Sq. Ft.)		
	Steep Slopes	Rolling Country	Flat Country
Acres			
10	5.6	1.9	1.1
20	9.4	3.1	1.9
30	12.8	4.3	2.6
40	15.9	5.3	3.2
50	18.8	6.3	3.8
60	21.6	7.2	4.3
80	27	8.9	5.4
100	32	10.6	6.3
125	37	12.5	7.5
150	43	14	8.6
200	53	18	10.6
300	72	24	15
400	89	30	20
Square Miles			
1	127	42	25
2	214	71	43
3	290	97	58
4	359	120	72
5	425	141	85
7	548	183	109
10	716	239	143
15	970	323	194
20	1204	401	241
30	1630	543	326
50	2390	797	478
75	3240	1080	648
100	4020	1340	805

In the installation of concrete pipe culverts, the pipe is laid in a carefully prepared trench, curved at the bottom to fully support the pipe. Back-fill and road cushions must be carefully placed and compacted in layers so that the concentrated loads of vehicles will be distributed over a large area and not come directly over a small portion of the pipe.

For small box and arch culverts in stable soil, the side walls usually constitute sufficient foundation bearing. In soft or doubtful soil and for large sized culverts, spread footings should be placed under side walls. Often the culvert floor is considered as the foundation footing. In such a case the floor, which acts as a beam, should be reinforced in the same manner as the culvert top, except that the steel is placed in the upper part of the slab.

All flat slab or box culverts, regardless of size, should be reinforced. As a rule, such reinforcing is placed with its center point 1½ inches from the bottom of the slab, except in the case of the floor slab, when it would be 1½ inches from the top. Reinforcing should be bent down and up into side walls a suitable distance. Care should be taken to see that it is held the required distance from the forms by metal spacers or other devices so that it will be in correct position in the concrete when the work is finished.

Wing Walls

Wing walls are provided on bridges and culverts to retain the road fill and prevent stream erosion. Such walls, used in connection with concrete pipe culverts, are generally built straight and parallel with the road. The top thickness of unreinforced end walls for pipe culverts should be not less than 12 inches and as a general rule the thickness at the bottom should be 0.4 the height of the wall. The foundation footing under the wall is usually made 6 to 12 inches wider than the wall.

End and wing walls for box or arch culverts are either straight and parallel with the road or flared at an angle to it. The flare type is more effective in confining the road fill. Especially should it be used on the upstream end of the culvert. End and wing walls are frequently reinforced in the interest of economy of concrete.

The concrete floor built in the bottom of concrete culverts assists in preventing choking of the waterway and undermining of the foundation. A vertical cut-off wall at each end of the floor, extending down 2 feet, is added protection against undermining. For very small culverts, the floor is made continuous with the walls and thus acts practically as a foundation. In larger culverts the floor is laid usually as a 6-inch pavement between the walls in order to properly distribute concentrated loads.

Road covering over all culverts should be 2 feet for dirt or macadam. Owing to the better distribution of loads by concrete pavement, it will be sufficient if the road slab proper is laid directly on the top of the culvert. This allows a greater clear height to the waterway, without altering the grade of the road, than when the dirt or macadam road cushion must be provided.

Since the upper part of the culvert top slab is acting in compression, it should not be made part of the concrete road. The additional thickness of the road slab should be laid upon the culvert top. In doing this, it is best to paint the top of the culvert slab with something that will prevent bond between it and the road slab. The road section over the culvert should extend some distance on each side of the culvert and be reinforced. In this way no joint is formed at the side of the culvert and danger of settling of the road slab at sides is avoided.

Concrete culverts are made the full width of the road, including the shoulders. This is done by widening the road slab to the full length of the culvert.

Bridges

Most of the foregoing notes on concrete culverts apply to concrete bridges. A bridge implies a larger structure, although there is no line drawn to definitely state where the structure changes from the classification of culvert to that of bridge.

The type of highway bridge to be selected for any given location depends upon length of span, waterway opening required, foundation conditions, etc. The flat slab bridge is probably the simplest form and is specially adapted to short spans and to the loads of ordinary highway traffic. It is merely a concrete slab of proper thickness, suitably reinforced and resting on abutments. Reinforcement usually consists of steel rods and requires little or no fabrication before placing. Form work also is simple. Essentially all that is required is a temporary wooden floor upon which the concrete may be supported until it has hardened sufficiently to carry its own weight and that of traffic.

Where loads and length of span are such as to require a heavier floor slab than common in the ordinary type of flat slab bridge, the design changes to a combination of girder and slab or girder, beam and slab. Girders extend parallel to the road between abutments and support the floor slab which can then be made relatively thinner. As the entire structure is a monolith, concrete beams act as "L" or "T" beams of which the floor slab is the upper flange. This results in economy by reducing the required size of beams. The beams, of course, are suitably reinforced. Sometimes on long spans, large concrete girders are used at each side with cross beams between, supporting the floor slab. Reinforcement in this type of structure is more complicated than in the types previously described.

In order to simplify the problem of reinforcing and of form construction, steel "I" beams are sometimes substituted for concrete beams. The "I" beam should be encased in concrete to prevent corrosion. The concrete floor slab rests on the "I" beams. In this type of construction it is convenient to hang the forms to the lower flange and thus eliminate the necessity of post supports for forms.

Concrete arches are suitable for locations where good foundation and sufficient head-room exist. Under such conditions the arch is economical. Form work is more difficult and extreme care is necessary to secure an unyielding foundation for the abutments. Both plain and reinforced concrete is used for highway arch bridges. In many cases the arch can be designed so that it will not need reinforcing. In other cases designing it for reinforcing will result in desirable economy of concrete.

Functions of Bridge Abutments

Bridge abutments have two functions. They help to support the bridge superstructure and to retain the fill of the approach. They may be either of plain or reinforced concrete. If reinforced, they allow considerable reduction in the abutment thickness and a consequent saving in concrete.

Wing walls are needed for all highway bridges, regardless of size. They prevent undermining and erosion of the abutment and also add to the stability of the abutment, as well as help to retain the fill of the road and bridge approaches.

In the building of abutments, retaining walls and wing walls, drainage should be taken care of by depositing against the back of the wall, a layer of broken stone for the full height, connecting this with weep holes or pipe extending through the wall.

Concrete for the ring of arch bridges may be placed in two ways unless the structure is so small that concreting can be finished in one day. The first method is to concrete a section of sufficient width so that the ring can be completed from abutment to abutment in one continuous operation. One section is completed each day or during a given period, until the full width of the bridge is reached. The second method is to deposit the concrete in transverse sections, extending the full width of the bridge in a manner similar to that after which masonry arches are built. In this method, the surface of the concrete at the end of each day's work must be left in a plane perpendicular to the form centering at that point. In the larger arches built by the latter method, there is some danger that the placing of concrete at the haunches will cause the centering to rise at the crown, thus distorting the true curve of the arch. This danger is slight in small arches, but can be guarded against by placing the keystone section and the haunch at the same time, the weight of the first tending to hold down the centering. In placing concrete after either method, it should be deposited first at both haunches and in equal amounts on each side so as to load the forms equally.

Foundation Material of Utmost Importance

The stability of arch bridges depends in a great measure on unyielding abutments and foundations. The character of the foundation material is therefore of utmost importance. Rock, hardpan, and hard and compact gravel, are suitable arch abutment foundations. Piles are required when the soil is yielding. Foundation footings must be carried down sufficiently to prevent underscouring or undermining by the current. Where necessary, they should be protected by rip-rap. For very large mass work it is permissible to use field stones in the concrete to effect a saving of materials. These should not be thrown in the forms indiscriminately, but be distributed carefully by hand during the placing of the regular concrete mixture, so that they will be evenly scattered throughout the mass.

In order to prevent seepage of water through the joints in the arch ring, the ring must be thoroughly waterproofed. While well made concrete in itself is sufficiently waterproof for this purpose, nevertheless there are joints at the end of each day's work which might allow seepage that would detract from the appearance of the arch. To prevent this the back of the arch and inner faces of the spandrel walls are given a ⅛-inch coating of coal tar pitch. To dispose of surface water, broken stone drains are provided to lead it away from the abutment or to drain pipes extending through the haunch of the arch near the springing line. The road fill may be of broken stone, cinders or earth. A minimum thickness of 2-feet of fill, well compacted, should be provided at the crown.

If a permanent hard surfaced road is to be installed over the bridge, it is good policy to wait for at least a year so that all settlement of the fill may have subsided.

Carpenter Work on Bridge

Carpenter work required on average bridge forms can easily be done by an ordinary carpenter. It should be borne in mind that various members of bridges are of considerable size and weight, therefore, forms should be strong and rigid to sustain the heavy loads imposed by the fresh concrete. In order that dimensions and shape of all members, when finished, will conform exactly with the design, every precaution should be taken to prevent distortion of forms. Extra care expended on making and erecting them will be repaid by the better appearance of the finished work. Where necessary, plenty of posts, well braced, should be used to support forms and prevent them from sagging with the green concrete.

The centering for an arch bridge is necessarily more elaborate than for a simple flat bridge, yet the carpenter work is well within the abilities of the carpenter of average skill. Special attention must be given to the wedges which provide for easy and gradual lowering of the centering. Only good quality lumber should be used for arch centering, particularly for main supporting timbers and bracing. Unless under exceptional circumstances, forms should never be lowered from beneath arch rings or floor systems in less than one month from the placing of the last concrete. If the work has been done under unfavorable conditions of temperature, a longer time even may be necessary.

Table No. 14

TABLE SHOWING QUANTITIES OF MATERIALS REQUIRED IN CONCRETE BRIDGES OF SPANS 8 FEET TO 24 FEET, ROAD-WAY 20 FEET, AS SHOWN BY THE STANDARD PLANS OF THE WISCONSIN HIGHWAY COMMISSION

Size	Excavation	Concrete	Reinf. Steel Pounds
8-foot span, 20-foot roadway	25 cu. yd.	42.3 cu. yd.	1910
10-foot span, 20-foot roadway	30 cu. yd.	49.8 cu. yd.	2210
12-foot span, 20-foot roadway	35 cu. yd.	56.5 cu. yd.	2720
14-foot span, 20-foot roadway	35 cu. yd.	60.3 cu. yd.	3100
16-foot span, 20-foot roadway	40 cu. yd.	68.6 cu. yd.	3540
18-foot span, 20-foot roadway	40 cu. yd.	72.8 cu. yd.	3950
20-foot span, 20-foot roadway	45 cu. yd.	81.4 cu. yd.	4540
22-foot span, 20-foot roadway	50 cu. yd.	91.9 cu. yd.	5320
24-foot span, 20-foot roadway	50 cu. yd.	98.2 cu. yd.	6360

CHAPTER 10.

CONVENIENT ESTIMATING TABLES AND EXAMPLES OF USE

For convenience, concrete is usually mixed in batches, each requiring one or more sacks of cement. The following table shows the cubic feet of sand and pebbles (or crushed stone) to be mixed with one sack of cement to secure mixtures of the different proportions indicated in the first column. The last column gives the resulting volume in cubic feet of compacted mortar or concrete.

Table No. 15

Mixtures			Materials			Concrete cu. Ft	
Cement	Sand	Pebbles or Stone	cement in Sacks	Sand Cu.Ft.	Pebbles or Stone Cu. Ft.	Mortar	Concrete
1	1.5		1	1.5		1.75	
1	2		1	2		2.1	
1	3		1	3		2.8	
1	1.5	3	1	1.5	3		3.5
1	2	3	1	2	3		3.9
1	2	4	1	2	4		4.5
1	2.5	4	1	2.5	4		4.8
1	2.5	5	1	2.5	5		5.4
1	3	5	1	3	5		5.8

The following table gives the number of sacks of cement and cubic feet of sand and pebbles (or stone) required to make one cubic yard (twenty-seven cubic feet) of compacted concrete proportioned as indicated in first column.

Table No. 16

Mixtures			Quantities of Materials		
Cement	Sand	Pebbles or Stone	Cement in Sacks	Sand Cu. Ft.	Stone or Pebbles Cu. Ft.
1	1.5		15.5	23.2	
1	2		12.8	25.6	
1	3		9.6	28.8	
1	1.5	3	7.6	11.4	22.8
1	2	3	7	14	21
1	2	4	6	12	24
1	2.5	4	5.6	14	22.4
1	2.5	5	5	12.5	25
1	3	5	4.6	13.8	23
1	3	6	4.2	12.6	25.2

Example No. I

How much cement, sand, and pebbles will be required to build a feeding floor 30 feet by 24 feet, 5 inches thick?

Multiplying the area (30 by 24) by the thickness in feet gives 300 cubic feet, and dividing this by 27 gives 11-1/9 cubic yards as the required volume of concrete. A one-course floor should be of 1:2:3 mixture, see Table No. 1 page 28. Table No. 16 shows that each cubic yard of this mixture required 7 sacks of cement, 14 cubic feet of sand, and 21 cubic feet of gravel or stone. Multiplying these quantities by the number of cubic yards required (11-1/9 gives the quantities of material required, eliminating fractions) as 78 sacks of cement, 156 cubic feet of sand, and 233 cubic feet of pebbles or stone. As there are 4 sacks of cement in a barrel, and 27 cubic feet of sand or pebbles in a cubic yard, we shall need a little less than 20 barrels of cement, 6 cubic yards of sand, and 9 cubic yards of pebbles or stone.

Example No. II

How many fence posts 3 by 3 inches at the top, 5 by 5 inches at the bottom, and 7 feet long can be made from one sack of cement? How much sand and pebbles will be needed?

Fence posts should be of a 1:2:3 mixture. Table No. 15 page 135 shows the volume of a one-sack batch of this mixture to be 3-9/10 cubic feet. The volume of one concrete post, found by multiplying the length by the average width and breadth in feet (7 by $\frac{1}{3}$ by $\frac{1}{3}$) is 7/9 cubic feet. By dividing 3-9/10 by 7/9 we find that five posts can be made from 1 sack of cement when mixed with 2 cubic feet of sand and 3 cubic feet of pebbles.

Example III

What quantities of cement, sand and pebbles are necessary to make 100 unfaced concrete blocks, each 8 by 8 by 16 inches?

The product of height, width and thickness, all in feet ($\frac{2}{3}$ by $\frac{2}{3}$ by 4/3) gives 16/27 cubic feet as the contents of a solid block. As the air space is usually estimated as $33\frac{1}{3}$ per cent, the volume of concrete in one hollow block will be $\frac{2}{3}$ of 16/27 or 32/81 cubic foot; in 100 blocks the volume of concrete will be $\frac{3200}{81}$=39.5=1.46 cubic yards, or $66\frac{2}{3}$ cubic feet which being divided by 27 gives a little less than $1\frac{1}{2}$ cubic yards. Unfaced concrete block should be $1:2\frac{1}{2}:4$ mixture, see Table No. 1, page 28. Table No. 16 shows that each cubic yard of this mixture requires 5-6/10 sacks of cement, 14 cubic feet of sand, and 22-4/10 cubic feet of pebbles. Multiplying these quantities by the number of cubic yards required ($1\frac{1}{2}$) gives the quantities of material required as 8-2/5 sacks of cement, 21 cubic feet of sand, and 33-3/5 cubic feet of gravel.

Example IV

How many 6-foot hog troughs 12 inches wide and 10 inches high can be made from 1 barrel of cement?

Use a 1:2:3 mixture, see Table No. 1 page 28. Table No. 15 shows the volume of a 1-sack batch of this mixture to be 3-9/10 cubic feet. As there are 4 sacks in 1 barrel, a barrel of cement would be sufficient for four times 3-9/10, or 15-6/10 cubic feet of concrete. The product of the three dimensions, all in feet, gives the volume of one trough as

5 cubic feet. However, approximately 30 per cent of this volume is in the open water basin or inside of the tank, leaving 3-5/10 cubic feet as the solid contents of concrete in one trough. Dividing 15-6/10 by 3-5/10, we find that 4 troughs (and a fraction over) can be made from 1 barrel of cement when mixed with 8 cubic feet of sand and 12 cubic feet of pebbles.

Table No. 17

NUMBER OF SQUARE FEET OF WALL SURFACE COVERED PER SACK OF CEMENT, FOR DIFFERENT PROPORTIONS AND VARYING THICKNESS OF PLASTERING

Proportions of Mixture	Materials			Total Thickness of Plaster				
				1/2-in.	3/4-in.	1-in.	1 1/4-in.	1 1/2-in.
	Sacks Cement	Cu. Ft. Sand	Bushels Hair*	Sq. Ft. Covered	Sq. Ft. Covered	Sq. Ft. Covered	Sq. Ft. Covered	Sq. Ft. Covered
1:1	1	1	1/8	33.0	22.0	16.5	13.2	11.0
1:1 1/2	1	1 1/2	1/8	42.0	28.0	21.0	16.0	14.0
1:2	1	2	1/8	50.4	33.6	25.2	20.1	16.8
1:2 1/2	1	2 1/2	1/8	59.4	39.6	29.7	23.7	19.8
1:3	1	3	1/8	67.8	45.2	33.9	27.1	21.6

*Used in scratch coat only.

Note:—These figures are based on average conditions and may vary 10 per cent either way, according to the quality of the sand used. No allowance is made for waste.

Table No. 18

MATERIALS REQUIRED FOR 100 SQ. FT. OF SURFACE FOR VARYING THICKNESS OF PLASTER

Proportions	1:1		1:2		1:2 1/2		1:3	
Thickness (in.)	C. (sacks)	Sd. (cu.yd.)	C. (sacks)	Sd. (cu.yd.)	C. (sacks)	Sd. (cu.yd.)	C. (sacks)	Sd. (cu. yd.)
3/8	2.2	0.08	1.5	0.11	1.3	0.12	1.1	0.13
1/2	3.0	0.11	2.0	0.15	1.7	0.16	1.5	0.17
3/4	4.5	0.16	2.9	0.22	2.5	0.23	2.2	0.25
1	6.0	0.22	3.9	0.29	3.3	0.31	3.0	0.33
1 1/4	7.5	0.27	4.9	0.36	4.2	0.39	3.7	0.41
1 1/2	9.0	0.33	5.9	0.43	5.1	0.47	4.5	0.50
1 3/4	10.5	0.39	6.9	0.50	6.0	0.56	5.4	0.60
2	12.0	0.45	7.9	0.58	6.9	0.64	6.2	0.69

If hydrated lime is used it should be added in amounts of from 5 to 10% by weight of the cement.

Hair is used in the scratch coat only in amounts of 1/8 bushel to 1 sack of cement.

These figures may vary 10% in either direction due to the character of the sand.

No allowance is made for waste.

Table No. 19

QUANTITY OF CEMENT REQUIRED PER CUBIC FOOT AND
PER CUBIC YARD OF CONCRETE FOR VARIOUS
MIXTURES IN TERMS OF SACKS AND BARRELS

1 Cu. Ft. Concrete	Sacks of Cement	1 Cu. Yd. Concrete	Bbl. of Cement
1:1:1	.5404	1:1:1	3.375
1:1½:3	.2808	1:1½:3	1.895
1:2:4	.2220	1:2:4	1.498
1:2½:5	.1848	1:2½:5	1.247
1:3:6	.1570	1:3:6	1.060

Table No. 20

MATERIALS REQUIRED FOR 100 SQ. FT. OF SURFACE
FOR VARYING THICKNESS OF COURSE

Thickness Mix	1 in.			2 in.			4 in.			5 in.		
	C.	Sd.	St.	C.	Sd.	St.	C.	Sd.	St.	C.	Sd.	St.
1:2	3.9	0.29	....	7.9	0.58	....	....	....	....	....	....	....
1:1:1	4.2	0.15	0.15	8.3	0.31	0.31	....	....	....	....	....	....
1:1:1½	3.7	0.14	0.20	7.3	0.27	0.41	....	....	....	....	....	....
1:1½:2½	2.6	0.14	0.24	5.1	0.28	0.47	....	....	....	....	....	....
1:1½:3	....	....	....	....	....	....	9.4	0.52	1.04	11.7	0.65	1.30
1:2:3	....	....	....	....	....	....	8.6	0.64	0.95	10.8	0.80	1.19
1:2:4	....	....	....	....	....	....	7.4	0.55	1.10	9.3	0.69	1.37
1:2½:4	....	....	....	....	....	....	6.9	0.64	1.02	8.6	0.80	1.27
1:2½:5	....	....	....	....	....	....	6.2	0.57	1.14	7.7	0.72	1.43
1:3:6	....	....	....	....	....	....	5.2	0.58	1.16	6.5	0.73	1.45

Table No. 20

MATERIALS REQUIRED FOR 100 SQ. FT. OF SURFACE
FOR VARYING THICKNESS OF COURSE

Thickness Mix	6 in.			7 in.			8 in.			9 in.		
	C.	Sd.	St.	C.	Sd.	St.	C.	Sd.	St.	C.	Sd.	St.
1:1½:3	14.0	0.78	1.56	16.4	0.91	1.82	18.7	1.04	2.08	21.1	1.17	2.34
1:2:3	12.9	0.95	1.43	15.0	1.11	1.67	17.2	1.27	1.90	19.3	1.43	2.14
1:2:4	11.1	0.82	1.64	12.9	0.96	1.92	14.8	1.10	2.19	16.7	1.23	2.47
1:2½:4	10.3	0.95	1.53	12.0	1.11	1.78	13.8	1.27	2.03	15.5	1.43	2.29
1:2½:5	9.2	0.86	1.72	10.8	1.00	2.00	12.3	1.14	2.29	13.9	1.29	2.57
1:3:6	7.9	0.87	1.74	9.2	1.02	2.03	10.5	1.16	2.32	11.8	1.31	2.61

Note:—Quantities expressed in the following units:
Cement........sacks   Sand........cubic yard
Pebbles or Broken Stone...............cubic yards

Table No. 21

MATERIALS REQUIRED FOR 100 SQ. FT. OF SIDEWALKS AND FLOORS FOR VARYING THICKNESS OF COURSE

Concrete Base

Proportions Thickness	1:2:3			1:2:4			1:2½:4			1:2½:5		
	C.	Sd.	St.	C.	Sd.	St.	C.	Sd.	St.	C.	Sd.	St.
2½ in.	5.4	0.40	0.60	4.6	0.34	0.68	4.3	0.40	0.63	3.9	0.36	0.72
3	6.5	0.48	0.72	5.6	0.41	0.82	5.2	0.48	0.77	4.6	0.43	0.86
3½	7.5	0.56	0.84	6.5	0.48	0.96	6.0	0.56	0.89	5.4	0.50	1.00
4	8.6	0.64	0.95	7.4	0.55	1.10	6.9	0.64	1.02	6.2	0.57	1.14
4½	9.7	0.72	1.07	8.3	0.62	1.23	7.7	0.72	1.14	6.9	0.64	1.28
5	10.8	0.80	1.19	9.3	0.69	1.37	8.6	0.80	1.27	7.7	0.71	1.43
5½	11.8	0.88	1.31	10.2	0.76	1.50	9.5	0.87	1.40	8.5	0.78	1.57
6	12.9	0.96	1.43	11.1	0.82	1.64	10.3	0.95	1.53	9.2	0.86	1.72

Wearing Course

Thickness Inches	1:1		1:1½		1:2	
	C.	Sd.	C.	Sd.	C.	Sd.
½	3.0	0.11	2.4	0.13	2.0	0.15
¾	4.5	0.16	3.6	0.19	2.9	0.22
1	6.0	0.22	4.8	0.26	3.9	0.29
1¼	7.5	0.27	6.0	0.33	4.9	0.36
1½	9.0	0.33	7.2	0.40	5.9	0.43
1¾	10.5	0.39	8.4	0.46	6.9	0.50
2	12.0	0.45	9.6	0.53	7.9	0.58

Note:—Quantities expressed in the following units:
Cement........sacks   Sand........cubic yards
Pebbles or Broken Stone..............cubic yards

CHAPTER 11.

FOUNDATIONS AND FOOTINGS

Before foundations and their footings are built, it is presupposed that suitable tests by borings and test loads have been made on the soil at the building site to determine its bearing value. In order to sustain the weight of the structure to be placed upon it, the foundations must be started on soil of sufficient bearing capacity to sustain the proposed superimposed load. Frequently to avoid excavation to unnecessary depth the proposed load is distributed over a greater area of soil by starting the foundation proper on a suitable footing.

When the excavation required for the building having basement extending 40 or 50 feet below grade passes through a firm stratum into softer material or where any settlement at all would be considerable, the foundations usually consist of wood or concrete piles driven close together or of concrete piers extending to a lower stratum of hard clay or to bed-rock. Where piles are driven close together and do not bear on bed-rock, the soil is compacted and skin friction together with slight bearing at the foot of the piles sustains the superimposed load without any material settlement. Concrete piers, on the other hand, do not depend on friction except when it is impossible to carry them down to bed-rock, in which case their sustaining power is usually increased by increasing the diameter of the lower section to form a bell-shaped footing which gives additional bearing area.

Building codes in various cities specify the maximum load allowed on clay or other soils. For clay, the allowable unit pressures go as high as 7,000 pounds per square foot for spread foundations. Concrete piers are usually proportioned to carry a load of 40,000 to 50,000 pounds per square foot at the top. The load generally used for wood or concrete piles seldom exceeds 20 tons per pile but it will be found that in most cases long piles, driven to refusal will carry safely 50 tons. Foundations and footings are rarely or never reinforced, dependence being placed on their mass for the required strength. There are, however, exceptions.

For small structures, it is frequently possible to place the concrete for the foundation in the excavated earth trench without using forms until ground level is reached. This is true when the earth is sufficiently firm to prevent caving of the sides of the trench. However, when concrete is placed under such conditions, workmen should be cautioned against running wheel-barrows too near the edge of the trench, thus resulting in earth dropping into freshly placed concrete and causing porous pockets in the mass. The best way is to lay planks alongside of the trench. Also, in spading or tamping concrete for foundations where it is being deposited in a trench without forms, care should be taken not to knock down fresh earth into the concrete for the reason above mentioned. Sometimes, as in the case of excavations for a basement or cellar, the concrete will need a form only on the inside, the earth wall of the excavation serving in this case as the outer form.

Table No. 22

BEARING POWER OF SOILS

	Supporting Power in Tons per Sq. Ft.
Rock—in thick layers, in natural bed	200
Clay—in thick beds, always dry	4
Clay—inthick beds, moderately dry	2
Clay—soft	1
Gravel and coarse sand, well cemented	8
Sand—compact and well cemented	4
Sand—clean and dry	2
Loam soils	0.5

CHAPTER 12.

WATERPROOFING OF CONCRETE

Poor Concrete Responsible for Popular Belief that Concrete is not Watertight

Much concrete construction that has been done would give many the impression that concrete could not resist the passage of water. This has given rise to discussion as to the waterproofness or watertightness of concrete.

Good Concrete Properly Mixed is Water Tight

Probably no concrete ever has been, nor ever will be made that does not contain a considerable percentage of voids. But for all practical purposes concrete can be made watertight. The simplest way of doing this is to so proportion well graded materials that voids will be reduced to the lowest possible minimum, and as existing, will not be connected with one another so as to result in continuous open channels through the mass.

The fact that concrete well made and properly placed is essentially a watertight material, is proved in many structures long used successfully as containers for water, oil or other liquids.

Thin sections of concrete are likely to contain small fissures that will permit seepage. Also such sections are almost invariably porous because of the difficulty in placing thin sections of concrete in forms to insure uniform density of the structure throughout. However, countless structures such as tanks, standpipes and other concrete receptacles for fluids stand today as evidence that concrete can be made watertight for all practical purposes.

Fundamentals for Watertightness

Primarily, several fundamentals govern the success or failure to attain watertight concrete. These fundamentals when observed or disregarded, as the case may be, contribute to the success or failure of the desired end. If mixtures are not properly proportioned, if the materials of which they are proportioned are not graded so as to reduce voids to the lowest possible limit, if mixtures are too dry or too wet, if, after placing, the concrete is not protected against too rapid drying out, the work will not be watertight. It is necessary, of course, that most uses of concrete in building construction shall result in work that is watertight or waterproof because basement walls, floors and roofs fail in part of their intended usefulness unless such an end is attained.

Reinforcing steel must also be prevented from rusting and this can only be done when the concrete covering it is essentially impermeable to water.

Systems for Increasing Watertightness

There are three principal methods or systems employed to increase the watertightness of concrete. They may be termed the "integral," "superficial" and "membrane" methods. The first consists in adding a material to the concrete when it is mixed. The superficial method consists of coating the concrete surface with a preparation that will adhere to it and remain attached. The membrane method consists of putting on the concrete a coating distinct from it. While this coating may adhere to the concrete, it will not crack if the concrete does, because of being a distinct and somewhat elastic membrane, usually strengthened by felt or other fiber cloth and impregnated with a mastic asphaltic or bituminous material.

The three methods just summarized have advantages as well as disadvantages. For example, it is impossible to use the integral method to prevent water from seeping through concrete after the work has been finished. In integral as well as in superficial methods of treatment, cracks developing in the concrete would make the possible waterproofing medium or method in these classes of no avail. In the membrane as well as the superficial method, care must be taken not to puncture the waterproofing coat.

The three methods of waterproofing mentioned are closely allied and there are various processes of waterproofing on the market that are about halfway between two of the methods. For example, when melted paraffin is applied to a concrete surface with a brush, as is sometimes done to increase watertightness, the treatment belongs to the superficial method, though the paraffin may penetrate the concrete so far as to be classed as an example of the integral method. Some paints may be classed as belonging to the superficial methods though they are elastic enough to bridge over very small cracks that may develop in the concrete and thus come within the membrane class.

There are a number of proprietary preparations on the market, the use of which is urged by the manufacturers in the interest of increasing the watertightness of concrete. Without approving or condemning the claimed merits of any of these preparations, it may be said that in no case will their use be successful unless in proportioning, mixing and placing the concrete and protecting it after placed, all of the fundamentals of good concreting practice are recognized.

One of the oldest processes of waterproofing concrete is known as the Sylvester process. It consists of applying powdered alum and soft soap to the concrete. These materials combine chemically to form compounds that are insoluble in water and fill the surface voids in the concrete with an insoluble, gelatinous mass. As a rule the Sylvester process is applied by painting on the concrete two separate solutions of alum and soap.

Asphalt and coal tar are used for waterproofing, particularly the outside of foundation or basement walls. They are applied hot with a mop. Several coatings are usually applied.

CHAPTER 13.

PLACING CONCRETE UNDER WATER

When concrete must be placed under water, the work should be done in such a manner that the ingredients of the mix will not separate. Either a tremie must be used—that is, a large pipe through which the concrete is carried to a point near the bottom of the water; or large buckets with hinged bottoms, that can be lowered and from which the concrete can be dumped with least disturbance.

When a tremie is used, the pipe is gradually withdrawn and thus elevated upward as the mass of concrete deposited is increased. The best results are obtained by mixing the concrete moderately dry, although when depositing by means of a pipe or chute, the concrete is often mixed very dry. The difficulty often encountered in placing concrete under water usually results from lack of care to prevent separation of materials. If the concrete is thrown into the water or even allowed to settle through it, separation of materials is unavoidable.

One common and inexpensive method is to provide a closed rectangular wood chute or a circular metal one. This is placed with one end extending into the water and to the foundation in such a manner as to prevent the concrete from flowing out while the chute is being filled. When entirely filled, it is raised slightly, thereby permitting the concrete to gradually distribute itself and at the same time permit additional concrete to be placed in the chute, so that at no time can water enter.

In extensive work, a closed bucket with hinged bottom is often used. In some cases concrete has been placed under water in coarse jute sacks which were lowered to the bottom of the foundation. This method, however, is not dependable since frequently there is poor bond between different parts of the foundation. When the concrete is to be deposited

147

from an airtight receptacle lowered into the water, it should be mixed dry enough so that when the gate or trap door of the bucket is opened, the material will be discharged in a mass.

Cofferdams should be sufficiently tight to prevent current of water through the pit and in other respects the water should be quiet. The surface of the concrete deposited must be kept as nearly level as possible to avoid the formation of pockets which will retain laitance and sediment. Where concrete is not deposited continuously, all sediment should be removed from the surface of the concrete by pumping or some other means before concreting is resumed. No mixture leaner than 1:2:4 should be used when concrete is deposited under water.

CHAPTER 14.

NOTES ON SILOS, COAL AND MATERIAL BINS AND GRAIN TANKS

As a rule all structures coming under the above heading are circular in shape, the exception being that sometimes coal pockets and other material bins may be of rectangular form. However, because of the development of circular forms for such construction, the occasions where rectangular bins are used are those largely governed by the location. The circular type of structure is in general more popular and becoming more common.

Certain essentials of construction are alike in all of the structures mentioned. Because of the pressure of contained contents all must be the subject of engineering design, principally to insure that the concrete will be sufficiently reinforced to withstand internal pressure.

Reinforcement for any of the structures mentioned may be either rods or some one of the several types of mesh fabric used as concrete reinforcement. To illustrate the method of determining the kind of reinforcement required for a monolithic concrete silo for example, reference is made to an accompanying table which shows the spacing of horizontal reinforcing bars for silos of various inside diameters. Assume a silo 14 feet inside diameter and 40 feet high as an example. For an inside diameter of 14 feet, the table specifies ½-inch round rods. The column at the extreme left of the table gives the distance from the top of the silo for intervals of 5 feet. As the silo is 40 feet high, run down the column to the line "35 to 40 feet," then across to the column under "14 feet diameter." This shows the spacing as 12 inches which means that there must be a horizontal ring of ½-inch steel every 12 inches. This spacing applies to the first 5 feet above the floor. For the next 5 feet the spacing changes to 14 inches. Spacing becomes greater as the top of the silo is approached and can be found by simply following the 14-foot column to the top. This method of determining the horizontal reinforcement applies to all sizes and heights of silos.

Table No. 23

SPACING OF HORIZONTAL REINFORCING RODS FOR SILOS OF VARIOUS INSIDE DIAMETERS

Distance in Feet Down from Top of Silo	10-foot Diameter ⅜-inch Round Rods	12-foot Diameter ⅜-inch Round Rods*	14-foot Diameter ½-inch Round Rods*	16-foot Diameter ½-inch Round Rods*	18-foot Diameter ½-inch Round Rods*	20-foot Diameter ½-inch Round Rods*
	Inch	Inch	Inch	Inch	Inch	Inch
Top 5 ft.	24	24	24	24	24	24
5 to 10	24	24	24	24	24	24
10 to 15	24	18	24	24	24	24
15 to 20	18	16	24	18	18	16
20 to 25	16	12	18	16	14	14
25 to 30	14	10	16	14	12	12
30 to 35	12	9	14	12	10	10
35 to 40	10	8	12	10	9	8
40 to 45	9	7	11	9	8	7½
45 to 50	8	6½	10	8½	7½	7

*If square rods are used increase spacing 30 per cent, but in no case should spacing be greater than 24 inches.

Vertical reinforcement is needed in all monolithic concrete silos. Usually this consists of ⅜ or ½-inch steel rods spaced 30 inches apart around the circumference of the silo regardless of its size. In this connection it should be mentioned that while theoretically the center of the wall is not the exact place where reinforcement should be placed in a circular structure, in designing for the reinforcement a factor of safety is considered which permits placing reinforcement at the center of the concrete section principally to facilitate placing of concrete. Attention also should be called to the necessity of securely wiring horizontal to vertical rods, correctly spaced at their intended location so that they may not be misplaced while depositing concrete.

Usually concrete of three different proportions is used in building monolithic silos. Requirements for materials are given elsewhere under "Aggregates." Table No. 1 page 28. A 1:2½:5 mixture is generally used for foundation and floor. For the walls a 1:2:4 mixture is used and for the roof a 1:2:3 mixture.

An accompanying table gives quantity of concrete materials for monolithic silos of various diameters.

Table No. 24

QUANTITY OF CONCRETE MATERIALS FOR MONOLITHIC SILOS OF VARIOUS DIAMETERS

These figures include footings and floor, but not roof. Walls 6 inches thick. Continuous doors 2 feet wide. Figures are for sacks of cement and cubic yards of sand and pebbles:

Inside Diameter Feet	For Silo 30 Feet High			For Each Additional 5 Feet in Height		
	Cement Sacks	Sand Cu. Yd.	Pebbles or Stone Cu. Yd	Cement Sacks	Sand Cu. Yd.	Pebbles or Stone Cu. Yd.
10	116	11	18	16.0	1.5	2.4
12	140	13	21.5	19.2	1.8	2.9
14	164	15	25	22.5	2.1	3.4
16	188	17.3	28.7	25.7	2.4	3.8
18	212	19.6	32.6	29.0	2.7	4.3
20	236	22	36.5	32.3	3.0	4.8

In order to assist contractors in giving information as to size of silo required, a table showing diameter of silo necessary to feed various numbers of animals and another showing the approximate capacity of round silos are given.

Table No. 25

DIAMETER OF SILOS REQUIRED TO FEED VARIOUS NUMBERS OF ANIMALS

Diameter in Feet	Approximate Minimum Pounds to be Fed Daily	Minimum number of each kind of stock to be fed from each size silo					
		Dairy Cows	Beef Cattle	Stock Cattle	500-lb. Calves	Horses	Sheep
10	525	13	21	26	44	48	75
12	755	19	30	38	63	69	252
14	1030	26	41	52	86	94	344
16	1340	34	54	67	112	122	446
18	1700	42	68	85	142	155	567
20	2100	53	84	105	175	191	700

Table No. 26

APPROXIMATE CAPACITY OF ROUND SILOS

Height of Silo Feet	Inside Diameter of Silo in Feet and Capacity in Tons					
	10 feet Tons	12 feet Tons	14 feet Tons	16 feet Tons	18 feet Tons	20 feet Tons
28	42	61	83			
30	47	67	91			
32	.51	74	100	131		
34	56	80	109	143		
36	61	87	118	155	196	
38	66	94	128	167	212	
40	70	101	138	180	229	280
42		109	148	193	244	299
44		117	159	207	261	320
46			170	222	277	340
48				236	293	361
50					310	382

Bins for holding materials other than grain such as coal, sand, etc., are sometimes built under trestles and are filled by dumping into an opening in the top, or by bucket, belt or screw conveyors. They are emptied through gates at or near the bottom or by dippers and grab buckets. Others are filled and emptied in much the same manner as grain bins.

The average grain elevator provides for the following parts and operation:

1. A receiving shed where the grain is dumped from wagons into chute ending in a boot at the foot of the elevator. Usually a scale is provided to weigh the grain as received.

2. The elevator is an endless chain or belt with buckets which carry the grain to a head-house where it is delivered by spout to the bins. A belt conveyor is generally necessary where there is a group of several bins.

3. When removing grain from the bin for shipping, it is usually spouted from the bottom of the bins into the boot at the foot of the elevator and from there elevated to the working floor.

4. Elevators may or may not contain drying and cleaning machines and other special machinery.

Lengths of rods used as reinforcement are generally such that splices must be made. The only reliable splice is to lap the ends enough to develop bond. If the pieces are in contact there will be loss of bonding area, amounting in the case of square bars to 25 per cent. For efficient bond the concrete must everywhere completely surround the steel. Lapped ends should have a clear space between them of not less than twice the thickness of the steel, the minimum distance being 1 inch regardless of the thickness of the steel. This may be obtained by placing a piece of steel or concrete between the ends of the rods and wrapping around them some soft wire to keep the ends separated while concrete is being placed. Horizontal reinforcement is generally placed on the outside of vertical reinforcement, merely because this is the easiest way to place it.

Accompanying tables show quantity of reinforcement required for bins of various diameter and height, and capacity of circular bins and tanks in bushels.

This table and the two following are based on data contained in the text book known as "Walls, Bins and Grain Elevators" by Milo S. Ketchum, published by the McGraw-Hill Book Co., New York.

Figures given for quantities of reinforcement required are theoretically correct but each bin or tank should be the subject of special engineering design so that the particular requirements in question to be met will be accurately determined.

Circular Grain Tanks

Area of horizontal reinforcing steel in square inches per foot of depth, and thickness of wall in inches. Steel to be in center of wall.

Set ¾-inch rods vertically at approximately 6-foot centers and ½-inch rods between them at approximately 2-foot centers.

Table No. 27

CIRCULAR GRAIN TANKS—DIAMETER IN FEET

Depth in Feet	8	10	12	14	16	18	20	22
5	0.026	0.033	0.040	0.046	0.053	0.059	0.066	0.072
10	0.041	0.066	0.079	0.092	0.105	0.118	0.131	0.145
15	0.050	0.073	0.096	0.120	0.157	0.194	0.197	0.217
20	0.056	0.082	0.109	0.140	0.171	0.207	0.234	0.289
25	0.058	0.087	0.119	0.153	0.190	0.228	0.268	0.309
30	0.059	0.090	0.125	0.163	0.204	0.246	0.294	0.336
35	0.060	0.091	0.128	0.170	0.213	0.260	0.309	0.360
40	0.062	0.092	0.131	0.174	0.219	0.270	0.324	0.381
45	0.064	0.094	0.134	0.179	0.225	0.288	0.335	0.398
50	0.066	0.095	0.135	0.183	0.235	0.290	0.347	0.410
Thickness of wall	6″	6″	6″	6″	7″	7″	7″	8″

Square Grain Tanks

Upper figures give thickness of wall in inches. Lower figures give area of reinforcement in square inch per foot of depth. Center of horizontal steel to be 1¼-inch from face of wall. Vertical bars ⅝-inch round or ½-inch square to be spaced 24-inch centers (approx.)

Table No. 28

SQUARE GRAIN TANKS—DIMENSIONS IN FEET

Depth in Feet	8'x8'	10'x10'	12'x12'	14'x14'	16'x16'	18'x18'	20'x20'	22'x22'
5	4″	4″	4.5″	6.0″	6.0″	6.5″	8.0″	8.5″
	0.23	0.28	0.34	0.43	0.50	0.55	0.60	0.65
10	4″	5.5″	6.5″	8.0″	9.0″	10.0″	10.5″	10.5″
	0.28	0.39	0.47	0.60	0.68	0.77	0.83	0.91
15	4.5″	5.5″	7.0″	9.5″	10.5″	11.5″	12.5″	14.0″
	0.31	0.42	0.52	0.74	0.86	0.94	1.02	1.15
20	4.5″	6.0″	7.0″	9.5″	11.0″	13.0″	14.0″	15.5″
	0.32	0.44	0.55	0.74	0.90	1.06	1.18	1.28
25	5″	5″	7.5″	10.0″	11.5″	13.0″	14.5″	16.0″
	0.34	0.46	0.58	0.80	0.93	1.06	1.22	1.34
30	5″	6.5″	8.0″	10.0″	11.5″	13.5′	15.0″	16.5″
	0.34	0.46	0.63	0.80	0.96	1.11	1.27	1.38
35	5″	6.5″	8.5′	10.5″	12.0″	14.0″	15.5′	17.0″
	0.35	0.48	0.66	0.82	0.97	1.14	1.27	1.44
40	5″	6.5″	8.5″	10.5″	12.0″	14.0″	15.5″	17.5″
	0.35	0.48	0.67	0.83	0.99	1.16	1.30	1.48
45	5″	6.5″	8.5″	11.0″	12.5″	14.0″	16.0″	18.0″
	0.35	0.48	0.68	0.84	1.00	1.18	1.33	1.51
50	5′	6.5″	8.5″	11.0′	12.5″	14.5″	16.0″	18.0″
	0.35	0.48	0.68	0.85	1.03	1.20	1.35	1.54

Table No. 29

CAPACITY OF CIRCULAR GRAIN BINS AND TANKS—IN BUSHELS

Height	Diameter in Feet							
	10	12	14	16	18	20	22	24
10	631	910	1238	1616	2042	2525	3060	3550
15	946	1364	1855	2420	3060	3785	4590	5320
20	1212	1820	2475	3230	4090	5050	6125	7100
25	1578	2275	3095	4040	5100	6310	7650	8880
30	1892	2730	3715	4840	6125	7575	9180	10630
35	2208	3185	4340	5650	7145	8840	10700	12400
40	2525	3640	4950	6460	8170	10018	12240	14560
45	2840	4095	5570	7270	9190	11350	13780	16380
50	3158	4550	6195	8080	10210	12620	15300	18200
55	....	5005	6814	8888	11231	13882	16830	20020
60	....	5460	7433	9696	12252	15144	18360	21840
65	....	....	8053	10504	13273	16406	19890	23660
70	....	....	8672	11312	14294	17668	21420	25480
75	....	....	9293	12120	15315	18930	22950	27300
80	....	....	....	12928	16336	20192	24480	29120
85	....	....	....	13736	17357	21454	26010	30940
90	....	....	....	14544	18379	22716	27510	32760
95	....	....	....	....	19399	23978	29070	34580
100	....	....	....	....	20420	25240	30600	36400

1 bushel = 2150 cu. inches = 1.245 cu.ft. or 1 cu. ft. = 1728 cu. inches = 1.245 cu. ft. per bu.

Note:—Special bins should always be the subject of special engineering design.

Table No. 30

TABLE OF WAGES

Rate per Hour

Hrs.	25	27½	30	32½	35	37½	40	42½	45	47½	50	52½	55	57½	60
1........	.25	.27½	.30	.32½	.35	.37½	.40	.42½	45	.47½	.50	52½	55	57½	.60
2........	.50	.55	.60	.65	.70	.75	.30	.85	.90	.95	1.00	1.05	1.10	1.15	1.20
3........	.75	.82½	.90	.97½	1.05	1.12½	1.20	1.27½	1.35	1.42½	1.50	1.57½	1.65	1 72½	1.80
4........	1.00	1.10	1.20	1.30	1.40	1.50	1.60	1.70	1.80	1.90	2.00	2.10	2.20	2.30	2.40
5........	1.25	1.37½	1.50	1.62½	1.75	1.87½	2.00	2.12½	2.25	2.37½	2.50	2.62½	2.75	2.87½	3.00
6........	1.50	1.65	1.80	1.95	2.10	2.25	2.40	2.55	2.70	2.85	3.00	3.15	3.30	3.45	3.60
7........	1.75	1.92½	2.10	2.27½	2.45	2.62½	2.80	2.97½	3.15	3.32½	3.50	3.67½	3.85	4.02½	4.20
8........	2.00	2.20	2.40	2.60	2.80	3.00	3.20	3.40	3.60	3.80	4.00	4.20	4.40	4.60	4.80
9........	2.25	2.47½	2.70	2.92½	3.15	3.37½	3.60	3.82½	4.05	4.27½	4.50	4.72½	4.95	5.17½	5.40
10........	2.50	2.75	3.00	3.25	3.50	3.75	4.00	4.25	4.50	4.75	5.00	5.25	5.50	5.75	6.00
20........	5.00	5.50½	6.00	6.50	7.50	7.50	8.00	8.50	9.05	9.50	10.00	10.50	11.00	11.50	12.00
30........	7.50	8.25	9.00	9.75	10.50	11.25	12.00	12.75	13.50	14.25	15.00	15.75	16.50	17.25	18.00
40........	10.00	11.00	12.00	13.00	14.00	15.00	16.00	17.00	18.00	19.00	20.00	21.00	22.00	23.00	24.00
50........	12.50	13.75	15.00	16.25	17.50	18.75	20.00	21.25	22.50	23.75	25.00	26.25	27.50	28.75	30.00
60........	15.00	16.50	18.00	19.50	21.00	22.50	24.00	25.50	27.00	28.50	30.00	31.50	33.00	34.50	36.00
70........	17.50	19.25	21.00	22.75	24.50	26.25	28.00	29.75	31.50	33.25	35.00	36.75	38.50	40.25	42.00

CHAPTER 15.

MECHANICAL EQUIPMENT—ITS STARTING CARE AND OPERATION

The best equipment that it is possible to make can be quickly ruined and its usefulness destroyed. by improper care and operation. Every contractor desires to get the best possible results from his equipment, and he hires the best obtainable operators to take care of his machines. At times, however, new men have to be broken in, and in order that these men may have the benefit of the experience of others, we have compiled in this chapter useful data as to the starting, care and operation of Koehring Mixers.

Embodied in this will be found data that will also be of interest and value to experienced operators. In addition, our Research and Engineering Departments will be glad to answer any questions submitted to them by operators, with a view to enabling them to secure the greatest possible output from their machines under existing conditions.

Method of Unloading Mixer

Assuming that the car containing a Koehring Mixer has arrived at the nearest siding to the work and that the blocking has been removed, jack up the mixer and put wheels, or multiplane tractors in place. A runway of at least 20 feet long should be built to make an easy grade. If an abundance of railroad ties is available, one may build a close cribbing of them, and use 3″ x 12″ planks for runways. If the ties are not to be had, heavier stringers may be hauled to the siding, as they do not require as much blocking as the above mentioned method. After the runway is built, block the wheels of the flat car securely, fasten block and tackle to the frame of the mixer and to the draw bar of the car on the opposite end from the cribbing. Use pinch bars under the truck wheels and work the mixer gradually onto the incline, lowering slowly by means of the blocks and tackle. A Paving Mixer may also be run off of car, and down runway under its own power, if desired.

Assembling of Paving Mixers

The first step in assembling a Koehring Paver, after unloading, is the placing of the overhead steel frame, which should be put together as indicated by the marks on the different parts. The tightening of bolts should be left until the framework is completely assembled as it may be necessary to adjust some parts of it by means of drift pins. By leaving bolts loose, the operation is made easier. After all bolts are in place, see that lock washers are on each, and then draw nuts up tight.

Next, load the charging skip of the paver on a truck, back the truck to the mixer and fasten skip in proper place. This method of attaching skip eliminates much lifting and blocking, and saves much time. The two skip cables are then led from the grooved winding sheaves, over the small sheaves near the top of the frame, to the cable brackets on the skip. Care should be taken to have the same tension on each cable to prevent twisting of the skip. It is also important that the cable ends are fastened securely.

Before the large sheave is keyed on, all cable should be run off and rewound to be sure that there is no twist in it. This is extremely important as the cable will continue to jump off the winding drum if twisted. Next, be sure that the sheave fits within 1/32" of the bearing; then drive the key home. When skip is on the ground the hoisting cable should practically fill the large sheave and there should be 1½ wraps on the winding drum. On the 14E Paver, in order that the cable may lead properly, the grooved winding sheave should be keyed on at the end of the shaft and not close up against the large sheave.

Next, load boom on truck, back truck to paver and secure boom in place. As the boom support cable is attached to boom, when shipped, it is a simple matter to hang it. Take cable off of sheave on boom where it is attached, and get the slack needed by turning the boom elevating hand wheel and then slip over sheave which is attached near top of frame.

On the 10E, 21E and 32E, the delivery bucket cable is reaved onto the grooved winding drum when shipped from the plant, so all that is necessary is to put it in the proper sheaves on the boom and attach to cable clamps on bucket as shown by blue print accompanying mixer.

On the 14E the cable must be re-reaved when assembled on the job, as the idler sheave is removed before mixer leaves the plant. To reave cable properly, stretch it on the ground in line with boom; take end nearest to mixer, lead it over top of idler or tension sheave on operator's side, bring it down in back of main winding grooved sheave and then up again to idler sheave. Repeat this operation until only two empty grooves remain on idler sheave, then skip a groove and bring it down in front of main winding sheave, then in front of rear upper sheave on boom, and over top of lower sheave. Then, take cable under boom and stop at bucket, which should be placed in center of boom. The other end of the cable is taken under the front sheave on top of boom, through the boom hoist bracket, over sheave on front end of boom, thence back to bucket. The ends are then taken around the shuttle and clamped in place. Be sure that cable is started in top center groove of shuttle and carried under the full length of shuttle before clamping.

Slack in bucket cable on the 10E and 14E is disposed of by taking up nuts on the idler or tension sheave hangers, and on the 21E and 32E by adjusting sheave on the end boom.

Boom Erection and Adjusting Instructions

Erect boom and place pin (1) in position. Then fasten support cable to support (2) and thread as per diagram and fasten other end at (3). Next slip assembled bucket and carriage on boom and bolt stop blocks (4) in position.

Boom Details

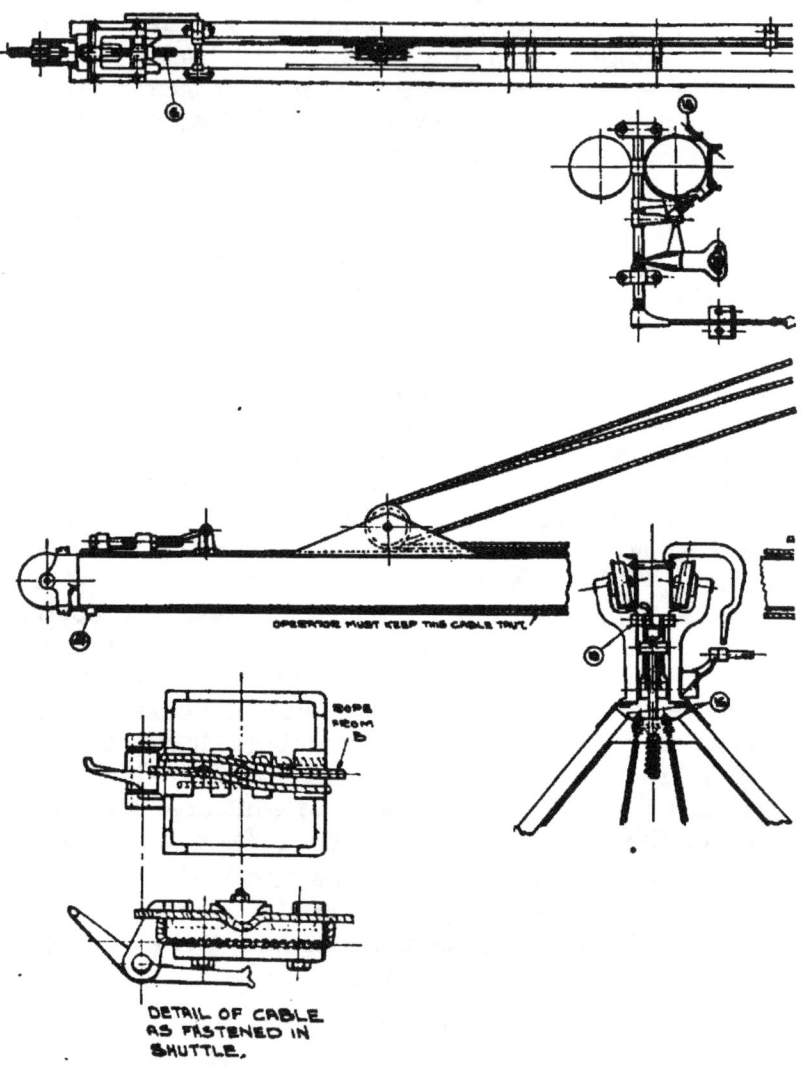

OPERATOR MUST KEEP THIS CABLE TAUT

ROPE FROM B

DETAIL OF CABLE
AS FASTENED IN
SHUTTLE.

Figure 1

Boom Details

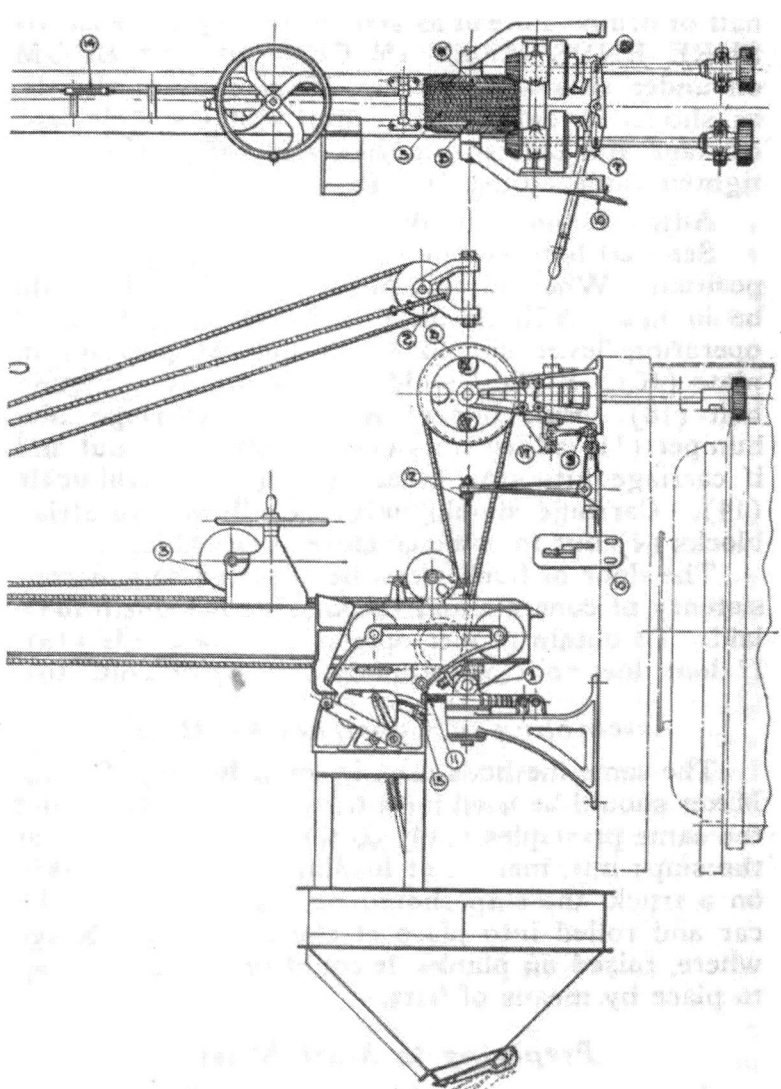

Figure 1

Boom Details

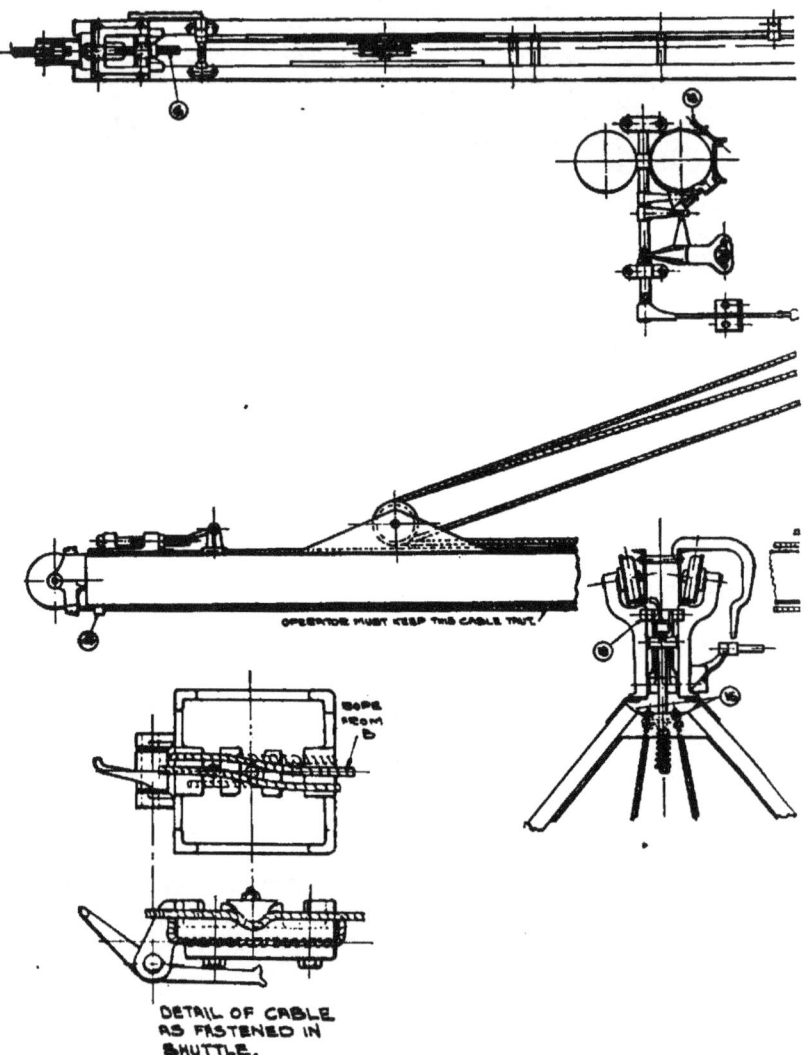

OPERATOR MUST KEEP THIS CABLE TRUE.

ROPE FROM B

DETAIL OF CABLE
AS FASTENED IN
SHUTTLE.

Figure 1

bricated. Be sure to keep all cables lubricated with good cable compound, and all gear teeth well lubricated with a good graphite grease, but DO NOT put grease on rollers or drum runway. The cable clamps should all be gone over to make sure all nuts are tight, as a loose clamp may cause trouble.

Before starting the engine, be sure that all clutches are disengaged; by so doing a bad accident may be avoided.

Starting and Care of Boiler

After mixer is assembled, a slow fire should be started in the boiler. Do not crowd the fire until steam pressure shows on the gauge, indicating that sufficient steam has been generated to protect the top ends of the tubes; which, in full length tube boilers are exposed to the hot gases. Hard firing in getting up steam is almost certain to damage the top ends of the tubes. This not only applies to firing the boiler the first time, but holds good for every time a new fire is made.

During the first day's operation, do not keep the water level in the boiler above the middle of the gauge glass, as the water will, undoubtedly, foam on account of the oil and grease on the plates. Blow down the water glass several times a day but never depend on it, as the opening may become choked, keeping the water level apparently constant, while the water in the boiler may actually be getting very low. The water cocks are put in the boiler to use, and are the only sure way of ascertaining the water level.

The boiler should be blown off well at the end of the day to remove all grease. After this, sufficient water should be carried in the boiler to show ¾ of the way up on the gauge glass.

Keep a thin clean fire. This will give you 100% more heat than a thick dirty fire, as the smoke from a thick fire chokes up the tubes, and combustion takes place not in the tubes but in the hood and stack. Clinkers should be removed from the fire box frequently and the ash box should be kept clean. Doing so prolongs the life of the grates and gives a better draft.

If at any time the boiler is to remain out of use for a few days, place a cover over the smoke stack to prevent water rusting out the tubes.

There will probably be injector trouble the first day, due to clogging from scales from new pipes and boiler. In such an event, take injector apart and clean thoroughly.

The boiler tubes should be kept clean, as the soot collected on them is a nonconductor of heat and more fuel is required in order to keep up steam. As often as is necessary, the tubes should be cleaned with a scraper—the frequency will depend entirely upon the fuel used and the carefulness exercised in firing; but they should be cleaned in the morning, at least, before firing up, and during the day they should be blown out every once in a while with steam.

Practically all feed waters contain more or less scale forming substances which precipitate and form incrustations in the water leg, on the flue sheet and around the lower end of the tubes.

Frequently the contractor has to make use of muddy water, and this also collects in the water leg and on the lower flue sheet. This scale and mud, if not removed by frequent cleansing, will become firmly baked on the heating surfaces, retarding the flow of the heat to the water and weakening the iron from stresses due to unequal expansion. An accumulation of scale 1/32″ thick requires 10% more fuel; 1/16″ of scale requires 20% more; ⅛″ of scale requires 30% more; and ¼″ requires 60% more. By keeping boilers clean, considerable fuel is saved.

Under usual conditions the boiler should be blown off a little every day. It is a good plan, before stopping after a day's run, to pump in more water than required while running. The next morning after firing is started and some ten to thirty pounds pressure has been raised, open the blow-off valve and blow the water down to the proper level. If the water is very muddy, it is a good plan to repeat this at the noon hour. After the boiler has been run for some length of time the boiler should be blown down entirely and thoroughly washed. This should be done at least once a week, and in case of muddy water, it could be done to advantage twice a week.

The boiler should not be blown down for washing while under steam pressure. The best time to do it is when the steam pressure has just gone down, but the water is still hot. Open the blow-off valve, let all the water run out, remove the handhole plates and wash out the boiler with a hose. To properly do this, it is necessary that the water be under pressure and that a properly shaped nozzle is used. A good nozzle can be made of $\frac{3}{8}''$ or $\frac{1}{2}''$ pipe having a short bend at the end so as to throw the stream of water at right angles to its length and at high velocity. If $\frac{1}{2}''$ pipe is used, the opening at the end of the pipe should be drawn down a little on a taper so as to give about a $\frac{3}{8}''$ opening.

The boiler should also be scraped with a scraper consisting of an elliptically shaped piece of iron, which would fit the side of the boiler, and fastened to a rod for handle. A very good cleaner can also be improvised from a heavy wire with a piece of chain secured to the end of it.

If the feed water contains ingredients such as to form a hard scale, impossible to remove by washing, a boiler compound may be used to advantage so as to reduce the scale to a muddy consistency that can be washed out.

The best water obtainable should always be used for the boiler. Where necessary to take it from a road side sump, it is well to make two sumps and use one as a settling basin; or barrels may be filled and water drawn from them after the sediment has settled to the bottom.

Starting and Care of Steam Engine

After all grease and oil cups and lubricator are filled, see that they are feeding properly. The oil cup on the connecting rod when full should last five hours; the lubricator should be adjusted to feed six drops per minute. The engine being oiled and enough steam in the boiler, see that throttle valve i‑

closed and drain cocks in cylinder and steam chest are open. Open valve in steam line near boiler, and the throttle valve just enough to blow out the water in steam line, steam chest and cylinder. After allowing the steam to pass through a few minutes, turn fly wheel off dead center, open throttle slowly until the engine starts running; then close drain cocks and open throttle valve wide. To avoid blowing out of gaskets, the nuts on the studs, holding cylinder head and steam chest cover in place, should be tested and tightened up, if necessary.

When starting an engine in cold weather, limber it up by letting it run idle a little while before placing any load upon it. When engine is started run it slowly, having all cocks open. Many cylinders are cracked by a sudden change in temperature.

In cold freezing weather, drain all water and oil from the cylinder and lubricator when engine is shut down for the night.

Gasoline Engine

For adjusting Fuller & Johnson Engine see their instruction book.

To Start and Run Fuller & Johnson Engine

1. Tighten the grease cups on main bearings and connecting rod and set feed on cylinder oil cup. Oil small parts with squirt can.

2. Fill jacket and water cooling tank with clean water. In the winter time the water should be warm to assist the engine in starting readily.

3. Fill the starting reservoir with gasoline and open the gasoline throttle to the starting mark (S).

4. Close the starting damper. (Handles in horizontal position as shown in Fig. 2).

5. Close switch on battery. (If battery ignition).

6. Prime the cylinder through the priming cup. Put in one-half to one priming cup full if the engine is cold. The hotter the engine the less gasoline required.

7. Attach the starting crank on the governor side and give the engine a few quick turns. As soon as the engine starts, open the starting damper and adjust the gasoline throttle to the running mark (R).

Line Drawing Showing Fuller & Johnson Gasoline Engine

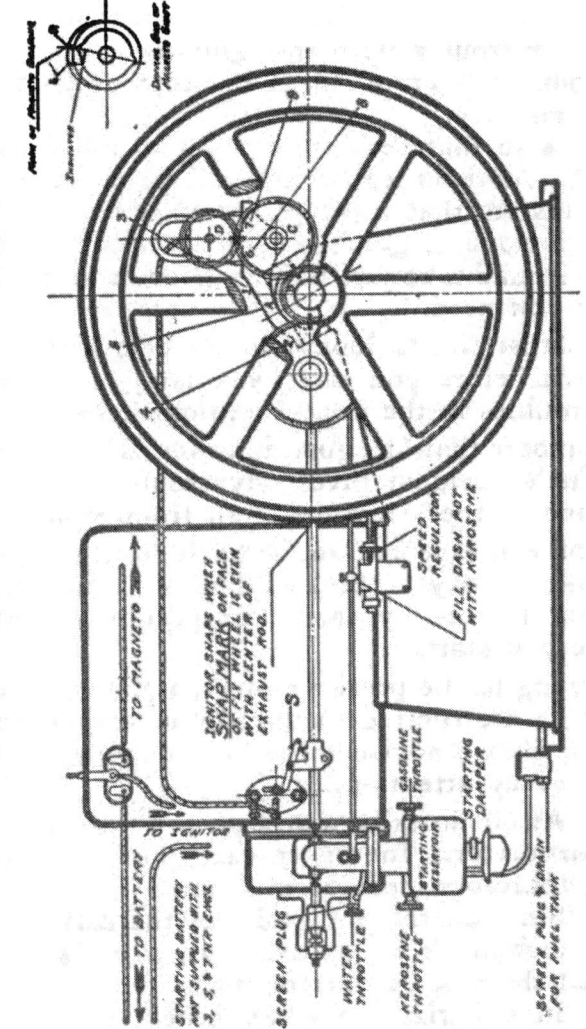

Figure 2

Care of and Starting Waukesha Motor

See that fuel tank is full.

Inspect the spark plugs to see that none is cracked or loose.

Test them for sparking. If you aren't getting a good spark at every plug, look for carbon on the plugs, or trouble with the ignition.

Look at your oil glass to see that your motor won't run dry.

Be sure that the cooling system is not short of water. An overheated motor will never give the good results that a perfectly cooled one will. Besides, if you are going where water is not right at hand, it may take you half an hour or so to get the water you need.

In other words, look your motor over from stem to stern before you start, so that you can correct any troubles in the easiest and quickest manner.

A motor that has good ignition, is well-oiled and well-fueled seldom gives any trouble—but neglect any one of these features, and troubles arise.

You can save yourself a lot of time, trouble, and expense—to say nothing of adding years to the life of your motor—by careful inspection of all parts before you start.

Owing to the presence of kerosene in some fuel, which is destructive to the motors if not properly vaporized, we advise that the following should be given every attention.

1. At all times be sure of a good adjustment on the carburetor. In ninety cases out of a hundred the carburetors feed too much fuel.

2. See that the air intake of the carburetor receives hot air from the exhaust, as it is most important that the carburetor bowls remain heated to assist in vaporizing the kerosene.

3. Although spark plugs cause little trouble these days, they should be removed at least once a week and have all the points uniformly adjusted no further apart than 1/32" of an inch. Guard the ignition wires; nothing tends to reduce the efficiency

of the motor more than poor ignition and carburetion. Make every spark do its work.

4. Do not overload the crank case with oil. Add oil several times a day and in this way retain a certain level at all times. Oil magneto once a week, putting about two drops of sperm oil in each oil hole with a match. Too much oil is as bad as not enough.

5. Drain the oil from the crank case at least once a week if using high test gasoline, or every third day when using the ordinary grade of gasoline. In doing this be sure to remove the four plugs under each connecting rod oil pocket as well as the large plug to the oil reservoir.

6. Keep the motor at its proper speed. Insist on having a seal placed on your governor, and never attempt to break it: it is one protection against motor troubles.

7. Watch the adjustment of push rods. Keep the valve seats in good condition. Any one cylinder working improperly will cause no end of trouble in a short time.

8. Cut down the idling of the motor, as doing this will reduce the dilution of the oil in the crank case, and the carbonizing of combustion chambers by at least 40%.

After the motor has been examined to see that the plugs are clean and that it has plenty of gasoline and oil, retard spark on magneto; throw out grounding switch; prime engine through the four priming cups, putting about two priming cups full of gasoline in each cylinder. Then the engine is ready to crank. When engine is running, advance spark. Before beginning to mix, see that there is proper water circulation by raising the return pipe in cooling tank.

To Get Out of Trouble

No matter whether you have a Fuller & Johnson engine, a Waukesha engine, or some other make, when trouble arises, consult the following TROUBLE CHART (reprinted by the courtesy of Stanton & Van Vliet Company, Chicago, Illinois, pu lishers of 'Gas Engine Troubles and How To Remedy Them' by J. B. Rathbun). Follow down the column under the head of SYMPTOMS until the description tallies with the actions of your engine.

Engine Will not Turn Over—Engine Stuck

1. Hot bearing or bearings seized.
2. Stuck piston due to overheated cylinder.
3. Water frozen to piston and cylinder (leaks).
4. Bolts rubbing on base or oil shields.
5. Friction clutch holding load on engine.
6. Broken gears wedging.
7. Water in cylinder due to leak in jacket.
8. Obstacle blocking wheels or gears.
9. Broken crank-shaft.
10. Dry bearings or rusted piston.

Starting Troubles

1. Fuel valve closed at tank (see that fuel reaches engine).
2. No gasoline in tank.
3. Battery or magneto switch open.
4. Broken or disconnected battery or magneto wire.
5. Dirty electrodes on make and break ignition system.
6. Broken igniter spring on make and break ignition system.
7. Weak batteries on either low or high tension system.
8. Magneto not generating on either low or high tension system.
9. Foul spark plug, high tension system.
10. Short circuit in wires or connection.
11. Defective spark coil on high tension system.
12. Defective timer.
13. Storage batteries.
14. No compression, indicated by engine turning easily over center.
15. Carburetor trouble due to poor mixture.
16. Cold weather carburetor troubles.
17. High altitude.

18. Air leaks in cylinder between carburetor and cylinder.

19. Vibrator on high tension spark coil may be out of action.

No Power—Loss of Power

1. Fuel valve partly closed, may jar shut.

2. Air damper closed in air intake pipe; always open the damper immediately after starting.

3. Compression relief cam may be left in "starting" position. It should always be turned to "run" as soon as engine is up to speed.

4. Throttle left in "starting" position.

5. Retarded spark will reduce power output; always advance it to the proper point as soon as engine is up to speed.

6. Advanced spark, when excessive, will cause a loss of power, which will be accompanied by heavy pounding.

7. Weak batteries reduce the spark and power.

8. Vibrator adjustment may be poor on high tension spark coil.

9. Foul igniters on make and break system.

10. Foul spark plugs are a frequent cause of power loss, especially with high compression.

11. Defective timer will cause power loss.

12. Misfiring is always accompanied by a loss of power.

13. Clogged muffler filled with soot, or a clogged exhaust pipe will cause power loss.

14. Long exhaust pipes or exhaust pipes with many short bends will reduce power.

15. Magneto trouble will reduce the spark.

16. Lack of oil, especially in the cylinder will cause compression leakage and power loss.

17. Hot bearings will cause the effects of an overload and reduce the output.

18. Carburetor troubles are a very frequent cause of power loss. "See Loss of Power."

19. Compression leakage is a very common and persistent cause of power loss. Test for compression by turning engine over "center" on compres-

sion stroke: if it passes over easily leakage exists, and must be stopped to prevent power loss and waste of fuel.

20. Cold jacket water will reduce power; have it leave jacket at 160 degrees F. on gasoline engines, and 200 degrees F. with kerosene engines.

21. Valve out of time will cause power loss.

22. Warm intake air will reduce mixture in cylinder.

23. High altitudes will reduce output of engine.

24. Worm cams, rollers and timing gears change timing.

25. Valve opening too small causes back pressure.

26. Spring too stiff on automatic intake valve.

27. Magneto out of time.

28. Valves stuck in guides.

29. Valve gear worn.

Misfiring

1. Loose wires or dirty connections.

2. Swinging ground caused by poor insulation.

3. Broken wire.

4. Weak or exhausted batteries.

5. Poorly adjusted vibrator on high tension system.

6. Foul spark plugs on high tension system.

7. Dirty electrodes on make and break system.

8. Moisture may cause short circuits.

9. Magnetos may cause misfiring.

10. Power loss is generally accompanied by misfiring.

11. Defective or short circuited spark coil.

12. Defective timer on high tension system.

13. Batteries may be weak.

14. Water in gasoline.

15. Valve gear worn.

16. Leaking exhaust valves are a common cause.

17. Poor mixture or poor carburetor adjustment.

18. Air leaks between carburetor and cylinder.

19. Valves out of time.

20. Leaking automatic intake valves.

21. Compression leaks.

22. Empty fuel tank.
23. Spark gap too large in spark plug.

Misfiring in One Cylinder

1. One cylinder may have a heavier carbon deposit.
2. One cylinder may have an air leak.
3. One cylinder out of time.
4. By timer having poor contact.
5. Loose wire leading to misfiring cylinder.
6. Sooted plug.
7. Magneto distributor foul with dust.
8. One vibrator stuck.

Sudden Stop

1. Ignition switch jarred open.
2. Fuel exhausted in tank.
3. Broken wire.
4. Loose connections or wires.
5. Carburetor nozzle clogged with dirt.
6. Fuel pipe clogged leading to carburetor.
7. Timer broken.
8. Defective magneto.
9. Hot bearings seize the shaft.
10. Defective igniter.
11. Water in gasoline.
12. Hot cylinder—Piston seized.
13. No oil.
14. Poor mixture.

Back-Firing

1. Poor mixture due to carburetor adjustment.
2. Retarded spark will cause back-fire.
3. Clogged carburetor nozzle or fuel pipe.
4. Leaky inlet valve on engine.
5. Air leaks in cylinder, or intake pipe.
6. Wide open throttle at full load.
7. On low speed may be caused by opening of the auxiliary air valve on the carburetor.
8. Valves out of time.
9. Defective timer on high tension system.
10. Weak batteries.

Irregular Running

1. Broken wire.
2. Dirty timer.
3. Sticking coil vibrator, high tension system.
4. Worn make and break mechanism (loose joints).
5. Loose timer control rods.
6. Water in gasoline.
7. Clogged carburetor nozzle.
8. Weak exhaust valve spring (broken springs).
9. Air leaks between carburetor and cylinder.
10. Worn cams or cam shaft on multi-cylinder engine as well as a twisted shaft or loose gears.

Overheating in Cylinder

1. Retarded spark.
2. Mixture too rich.
3. Lack of oil in the cylinder.
4. Poor water circulation due to the jacket.
5. Poor compression.
6. Insufficient valve lift.
7 Clogged exhaust pipe.
8. Clogged muffler.
9. Clogged radiator.
10. Defective circulating pump.
11. Tight piston.
12. Lime deposits in cylinder.
13. Overload on engine.
14. Closed water supply valve.

Pre-Ignition or Deep Pounding in Cylinder

1. Too much lubricating oil forms deposit.
2. Rich mixture forms a deposit in combustion chamber.
3. Overheated cylinder, especially in air cooled engines.
4. Sharp edges in combustion chamber.
5. Deposit in cylinder.
6. Deposits in kerosene engines are usually formed by running with a cold cylinder or by having the intake air too cold.

Smoke

1. Black smoke is caused by too much gasoline, or too rich a mixture.
2. Light colored smoke is caused by an excess of oil fed to the cylinder.

Engine Gradually Slows Down and Stops

1. Weak or exhausted batteries.
2. Poor mixture due to carburetor adjustment.
3. Overload on engine.
4. Magneto slipping or governor out of order.
5. Overheated bearings.

Excessive Vibration

1. Engine crankshaft may not be perfectly balanced.
2. Twisted cam shaft may change valve timing and cause an uneven application of power.
3. Uneven wear on cams or push rods may give an uneven application of power.
4. Defective timer may fire the cylinders at unequal periods.
5. Carburetor not situated at equal distances from the cylinders may be the cause of vibration.

Cam Shaft Rattle

1. Caused by retarded spark.
2. Loose cam shaft bearings or gears.
3. Loose cam rollers or pins.

Wheezing Scraping Sound

1. Broken piston rings.
2. Dry piston due to lack of oil.
3. Vibration of auxiliary air valve on carburetor.
4. Tight piston.
5. Overheated cylinder.
6. Fly wheel scraping on metal shields.

Knocking and Pounding—Regular

1. Ignition too far advanced.
2. Pre-ignition.
3. Overheated cylinders.
4. Loose bearings.

5. Loose connecting rod bearings.
6. Loose fly wheel.
7. Loose counterweights.
8. End play in crank shaft.
9. Broken valve stem.
10. Broken circulating pump.
11. Wear on cam shaft and cams.

Irregular Knock

1. Loose electrical connections.
2. Loose piping or rod on engine.
3. Pre-ignition.
4. Defective commutator or timer.

Speed Variation

1. Due to misfiring.
2. Water in fuel.
3. Irregular supply of gas.
4. Magneto slipping.
5. Defective fuel pump.
6. Defective governor.
7. Wear on valve gear.
8. Defective timer.
9. Loose electrical connections.
10. Poor mixture.

Adjustment of Friction Clutches

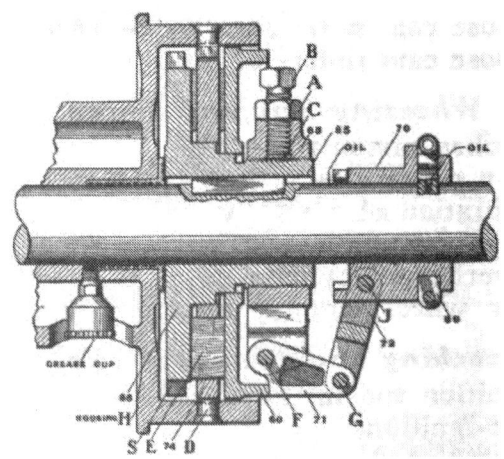

Figure 3

Unnecessary slipping of the friction clutch, causing undue wear on the friction blocks E should be remedied at once by making adjustments in the following manner:

1. Loosen lock-nut A on set screw B.
2. Loosen set screw B.
3. Strike set screw B lightly with a hammer to loosen copper disc C which has been forced into threads of master disc 65.
4. Turn adjusting ring 63 to the right about one-quarter of a turn to adjust for ordinary wear on blocks E. Try the clutch by pushing in spool 70 with lever or hand wheel. If the clutch should be too tight, i. e. does not lock, turn adjusting ring 63 to the left slightly.
5. When proper adjustment has been made TIGHTEN SET SCREW B, making sure that copper disc C is in place first, and then tighten lock-nut A.

CAUTION: Be sure that the fifth instruction has been complied with before operating the machine.

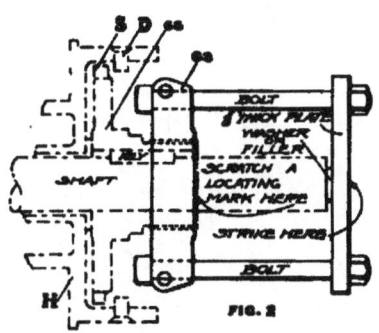

Figure 4

Removing Clutches

When for any reason it becomes necessary to remove a clutch, where it is impractical to remove the housing H first, proceed as follows:

1. Loosen lock-nut A on set screw B.
2. Loosen set screw B.

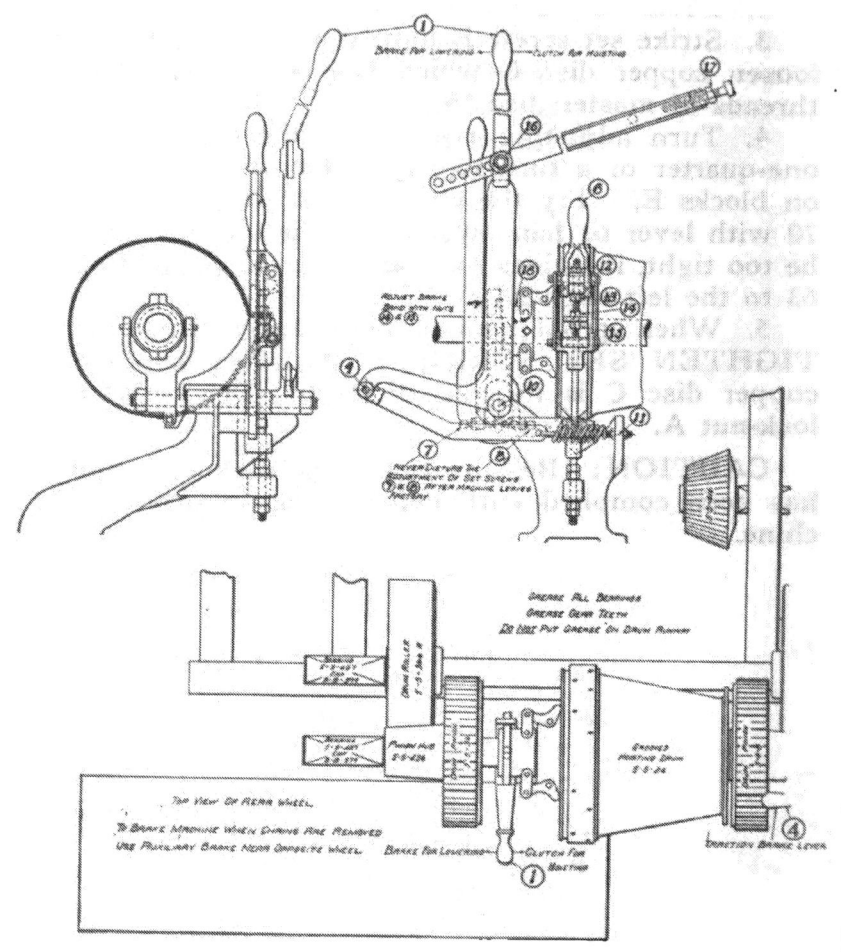

Figure 5

Main Shaft and Clutch Assembly 14E Paver

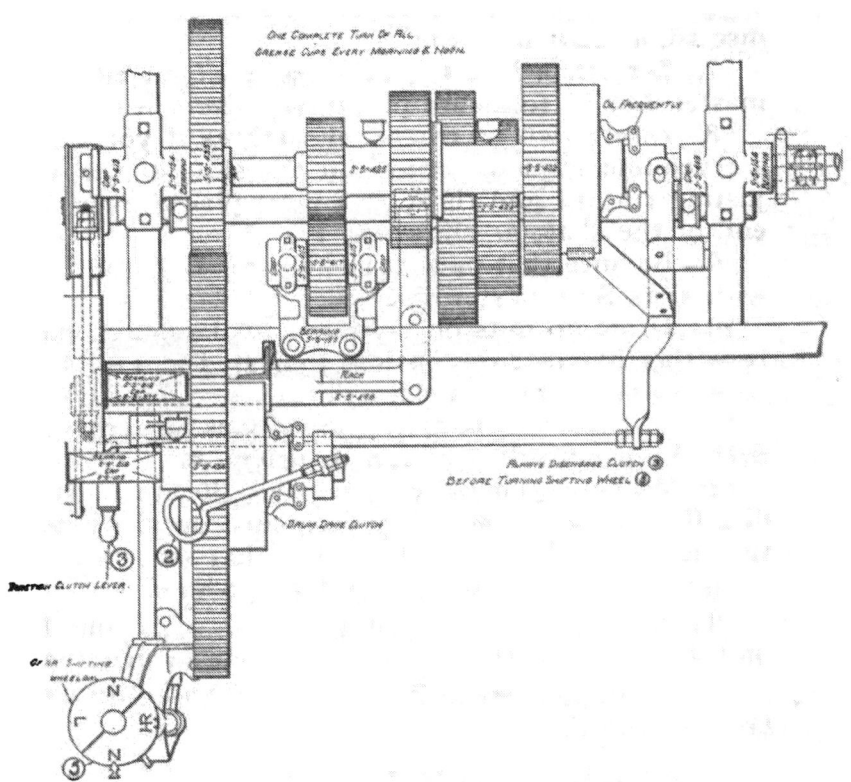

Figure 5

Main Shaft and Clutch Assembly 14E Paver.

3. Strike set screw B lightly with a hammer to loosen copper disc C, which has been forced into threads of master disc 65.

4. Unscrew adjusting ring 63 and remove same, together with toggle links 71, toggle yokes 72 and spool 70.

5. Remove from plate 60 and friction plate 74.

6. Remove pins F from adjusting ring and toggle links and replace adjusting ring 63 on master disc 65, as shown in Fig. 4.

7. Scratch a locating mark on shaft at edge of master disc 65 to insure proper replacement of disc.

8. Insert two or three bolts (three if you have a three-link adjusting ring) in the slots of the adjusting ring 63 and through a plate placed over the end of the shaft as shown in Fig. 4.

9. Be sure that pins D in housing H are in line with slots S in master disc 65.

10. Draw up nuts on bolts and strike occasionally with a hammer on plate as indicated in Fig. 4 in order to start the disc 65.

11. In cases where the housing can be removed first, the master disc 65 can be driven off.

12. When replacing clutch parts, locate master disc 65 by mark previously scratched on shaft, see that it is keyed tight and be sure that set-screw B is tightened before operating the machine.

Toggle links 71, toggle yokes 72 and pins F, and J should be replaced by new ones when they become worn to such extent as to cause improper operation of the clutch.

Adjustment of Hoist Brake

On all Koehring Mixers, the basic principle of adjustment is to first adjust the clutch, and then adjust the brake to suit. While the design of levers is somewhat different on the various mixers, the following general method of adjustment of the brake on the 14E and 21E Pavers applies to all the machines. For reference see Figure 5.

In case set screws (7) and (8) have been disturbed, after mixer has left the factory, to adjust properly—Pull lever (6) down and loosen nuts (14) and (15).

Adjust set screws (7) and (8) so that centers of (9), (10) and (11) are approximately in line when clutch (18) is engaged. Release clutch (18) with lever (1) and move back to extreme limit. Tighten nut (14) until tension pulls clutch (18) forward ¼", then tighten nut (15).

Readjust lever (6) by adjusting nut (13) until lever (6) locks brake band, then tighten nut (12).

After the brake is adjusted, adjust the knockout at (16) or (17).

Lever Operation

General:—

While the lever arrangement is somewhat different on the various mixers, the same operating principles apply to all machines, whether they be Pavers, Heavy Duty Mixers or Dandie Mixers.

As, on account of the propelling features, the lever arrangement is somewhat more complex on the pavers, it is described in detail:

10E Paver

Refer to Figure 6 page 182 for reference.

When starting mixer disengage gasoline engine clutch (not shown) or stop steam engine. Then disengage traction clutch with lever (1); shifting lever (2) must be in neutral position. Next, engage drum drive gear with lever (4) and engage gasoline engine clutch (not shown) or start steam engine.

To hoist charging skip, engage hoisting clutch with lever (6).

To stop hoisting of skip, disengage clutch with lever (6) and apply brake lever (7).

If, however, the skip is raised to charging height, it will be stopped automatically providing the adjusting screw is set properly in the knockout.

To lower bucket, release brake lever slightly.

To move mixer, hoist and lock charging skip clear of the ground, then:

(a) For low Speed Ahead—Lever (2) must be in neutral position.

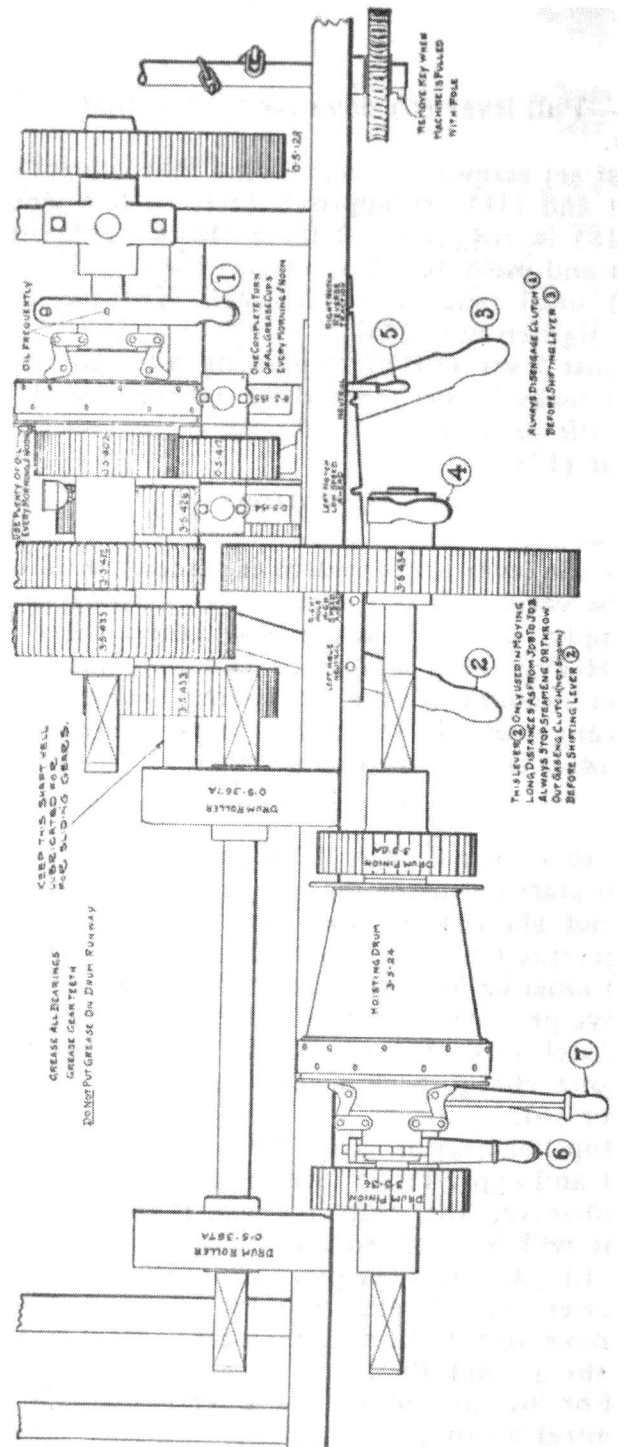

Figure 6

Gear Assembly and Control 10E Paver.

Disengage traction clutch with lever (1). Next, shift lever (3) to left hand notch. Then, engage clutch with lever (1).

To brake, throw out traction clutch with lever (1) and apply brake lever (5))

(b) For reverse traction—Disengage traction clutch with lever (1), having lever (2) in neutral position. Then, shift lever (3) to right hand notch and engage clutch with lever (1).

To brake, throw out traction clutch with lever (1) and apply brake lever (5).

(c) For high speed ahead—With gasoline engine clutch disengaged or steam engine stopped and lever (3) in neutral position, engage traction clutch lever (1) and throw out drum drive gear with lever (4). Then shift lever (2) from left hand hole to right hand hole and engage gasoline engine clutch or start steam engine.

To brake, make sure that traction clutch lever (1) is engaged. Then, throw out gasoline engine clutch or shut off steam and apply brake lever (5).

14E and 21E Pavers

For reference see Figure 7 pages 184 and 185.

To hoist skip, engage drum drive clutch with lever (2) then engage hoist clutch with lever (1).

To lower skip, disengage clutch lever (1) halfway and when skip nears ground apply brake with lever (1).

To hold mixer when operating on a grade, apply and lock brake lever (4).

To hoist batch box with derrick—Engage clutch (7) with lever (6); raise batch box high enough to clear skip, next apply brake with lever (6) swing batch box over skip and dump material. After batch box is empty, to lower box on car reverse operation.

To operate power discharge—Engage clutch (B) with lever (18) to discharge. To reverse swinging chute, engage clutch (A) with lever (8).

To move mixer, hoist skip to clear ground, then lock with lever (13).

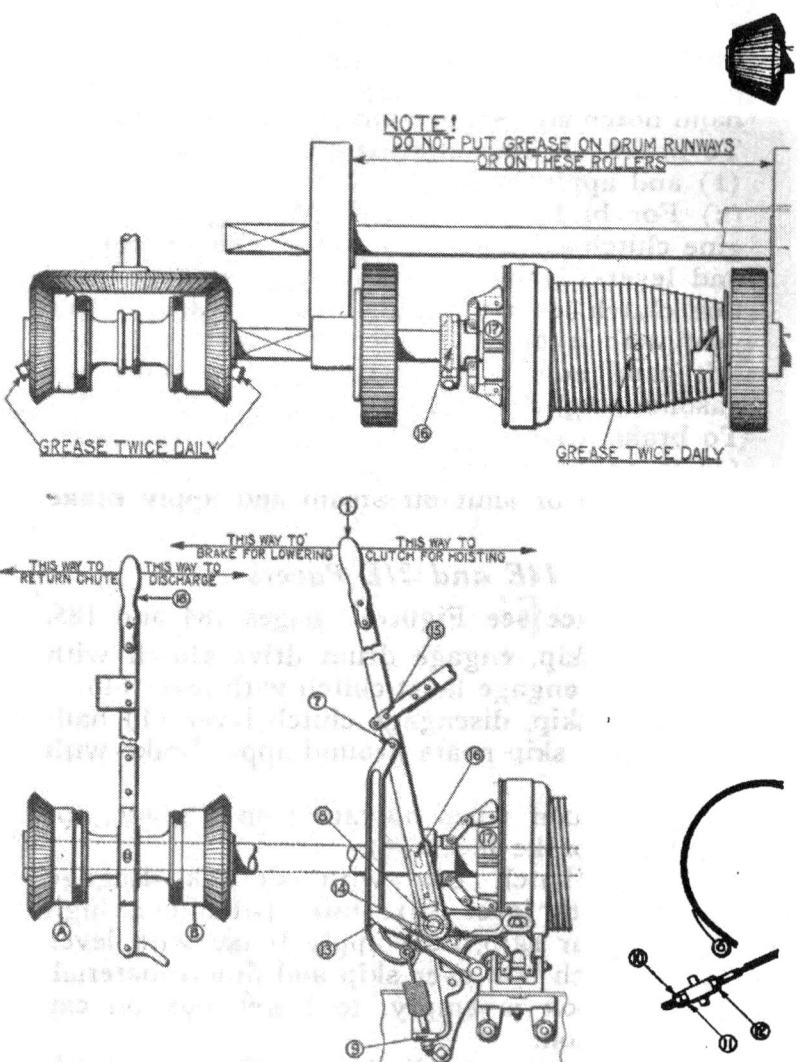

Figure 7

Main Shaft and Clutch Assembly 21E Paver.

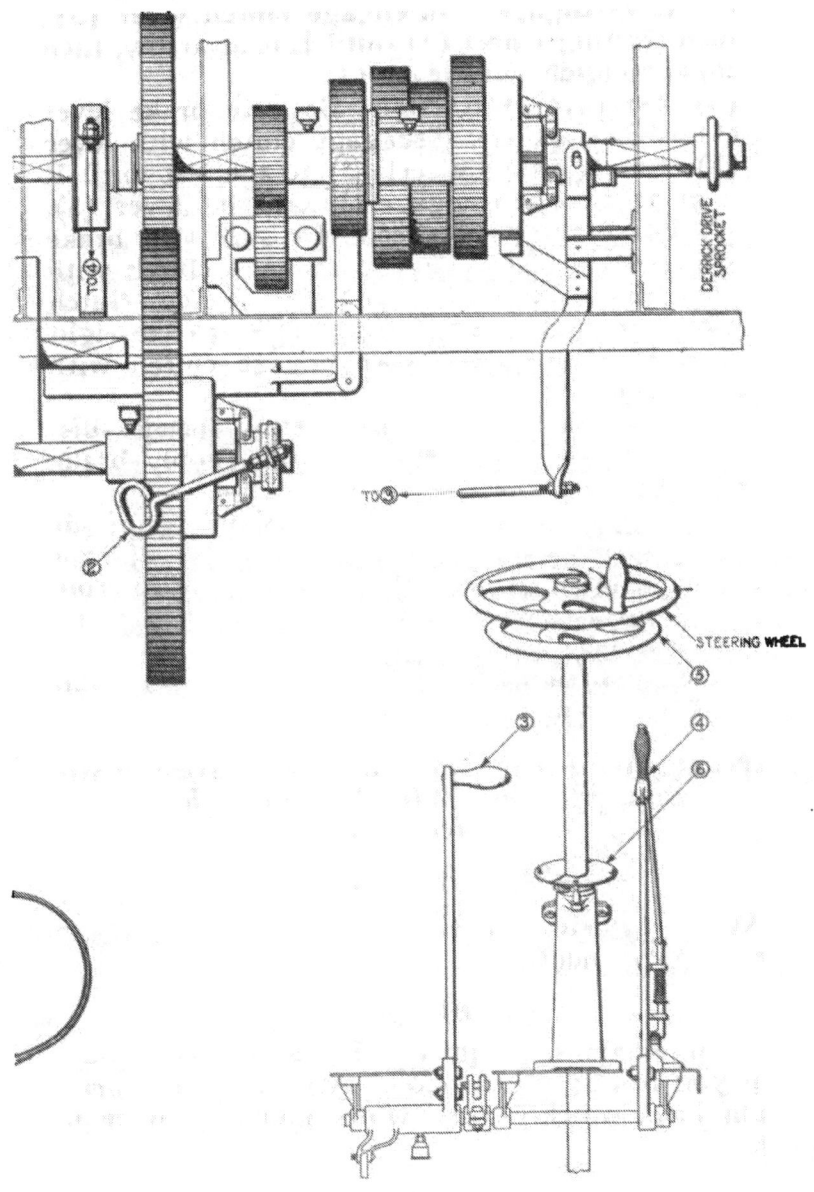

Figure 7
Main Shaft and Clutch Assembly 21E Paver.

(a) For slow speed forward, be sure brake lever (4) is disengaged; disengage clutch lever (3); turn shifting wheel (5) until L is at arrow, then engage clutch with lever (3).

(b) For reverse traction—Be sure brake lever (4) is disengaged; disengage clutch with lever (3); turn shifting wheel (5) to the left until R is at arrow, then engage clutch with lever (3).

(c) For high speed ahead—Be sure that brake lever (4) is disengaged; disengage clutch with lever (3); next disengage drum drive clutch with lever (2); turn shifting wheel (5) to right until H is at arrow, then engage clutch with lever (3).

(d) To work brake for all traction speeds—disengage clutch with lever (3) then apply brake lever (4).

When mixer is to be removed with teams or tractor instead of its own power, remove driving chains from rear wheels, disconnect steering rod, by removing cap from front ball socket, and be sure to wire the rod up well.

To apply brake under the above conditions, pull on brake lever near rear wheels.

Instructions for Operating Steering Mechanism on 21E Paver—With Full Length Multiplane

Straight Ahead

When traveling straight ahead leave steering wheel (A) in neutral.

Slight Turn

Turn wheel (A) quickly in desired direction until you feel a sudden stop. After proper course is obtained, quickly turn wheel (A) back to neutral.

Sharp Turn

Turn wheel (A) quickly in desired direction until you feel a sudden stop. Continue to turn wheel until no more movement can be obtained and hold firmly until proper course is obtained; then reverse wheel back to neutral.

Adjustments and Greasing

If brake drum sticks apply a few drops of oil on lining.

Grease cups (H) and (K) should be turned down freely twice a day; on a long travel 4 to 6 times daily.

If lever (B) goes over center without setting brake hard enough, loosen nut (D), remove pin (L), adjust clevis (E) until you obtain 1/16" clearance between brake shoe (G) and brake drum (F) on both drums.

Replace pin (L) and tighten nut (D).

Loosen nut (C), remove pins (M) and (N), adjust rod (P). When rod (P) is properly adjusted, lever (B) must be in neutral and 1/16 clearance maintained between each brake shoe and brake drum. Insert pins (M) and (N) and tighten nut (C).

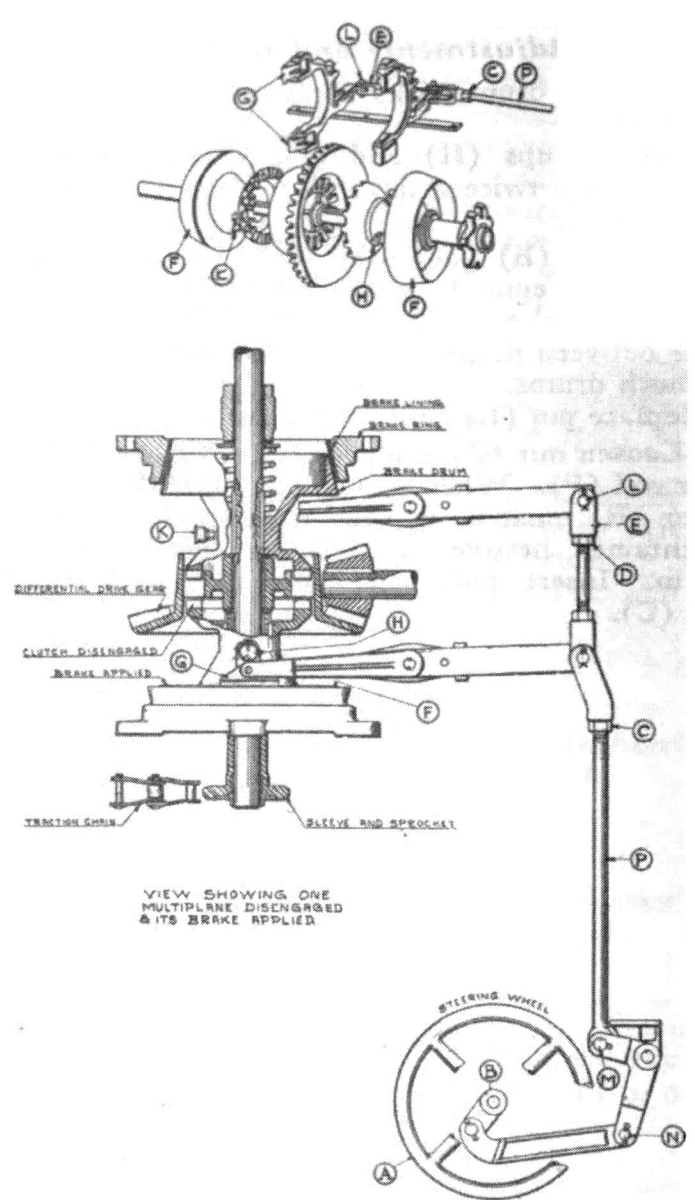

VIEW SHOWING ONE
MULTIPLANE DISENGAGED
& ITS BRAKE APPLIED

Steering Mechanism with Full Length Multiplane on 21E Paver.

Water Control Essential to Dominant Strength Concrete

Successful concrete construction is only possible when the quantity of mixing water is systematically controlled. Uniformity of water content in each batch means uniform strength of concrete. Without knowledge of the aggregates to be used, the weather conditions under which the work is to be carried on, and the amount of water in the sand, it is impossible to set a definite figure as the quantity of mixing water required per cubic foot or cubic yard of concrete. If a porous aggregate, such as crushed blast furnace slag, is employed, sufficient water must be added not only to hydrate the cement and wet the surface of the sand but to be absorbed by the coarse aggregate; while with a hard dense gravel less water is required.

With a given aggregate, known weather conditions and a particular type of work, there is one quantity of water which satisfies the requirements. This amount may change from day to day but will not change from one mix to the next.

Adjustment of Water Tank

Figure 8

In the design of the water measuring tank two requirements have been kept constantly in mind. First, that the measuring device be easily adjusted, and second, that the tank when set deliver the same quantity of water to each batch of concrete, measuring it automatically even though the operator be giving his attention to some other part of the mixing operation. Further, the foreman or inspector should be able at a glance to determine the quantity of water passing into the mixer.

The Koehring Water Measuring Tank fulfills these requirements. It is constructed without floats or intricate parts and can be quickly regulated and adjusted without the use of a wrench or special tools. Simple in design, staunch in construction, it is a dominant factor in the manufacture of Standardized Concrete.

The tank is filled by attaching a supply hose at nipple D, Fig. 8, and turning handle of three-way valve, into position C, Fig. 8. With the handle in this position, water will flow into the tank, the air escaping through check valve E, which automatically closes, when the tank is filled.

Discharge of water from the tank into the drum is accomplished by turning three-way valve handle into position C2, Fig. 8.

The following table gives the amount of water discharged from the tank for each setting of the regulating handle:

Settings of Valve Handle	Discharge from 12x24 Tank		Discharge from 12x33 Tank		Discharge from 16x34 Tank		Discharge from 18x40 Tank		Discharge from 18x50 Tank	
	Gal.	Lbs.	Gal.	Lbs.	Gal.	Lbs.	Gal.	Lbs.	Gal.	Lbs.
1	1.2	8.5	1.3	10.75	2.1	17.5	6.4	53.0	5.7	47.5
2	1.4	11.5	1.7	14.5	3.0	24.75	7.5	62.75	7.1	59.5
3	1.8	15.25	2.7	22.5	4.3	35.5	9.7	81.5	9.7	81.0
4	2.4	20.0	3.5	29.0	6.2	51.25	12.6	105.75	13.6	113.0
5	3.3	27.25	4.7	39.0	8.3	69.0	16.5	137.5	18.0	150.0
6	4.2	35.25	5.9	49.25	11.2	94.0	20.0	167.0	22.2	185.5
7	5.2	43.5	7.6	63.25	14.0	117.0	24.3	203.0	29.2	243.5
8	6.2	52.0	9.0	75.25	17.5	145.5	29.0	240.75	35.3	294.5
9	7.8	64.75	10.8	90.0	21.0	175.25	34.0	284.5	42.2	352.0
10	9.2	76.5	12.7	106.0	24.1	201.0	38.0	317.0	48.5	404.25
11	10.0	84.0	14.1	117.5	27.1	226.0	41.5	345.75	53.5	446.5
12	10.9	91.0	15.1	126.25	28.5	239.7	43.5	362.75	57.3	478.0

In starting up mixer, the segment on the handle B should be set with latch A in notches (4) or (5) and then moved backward or forward to regulate to the proper amount of water required.

If tank does not fill fast enough, do not blame the water measuring system but rather look for the trouble in the supply line. See if there is sufficient water pressure and be sure that the supply line or any part of it is not smaller than the supply opening at D.

The tank should be occasionally drained during the working season in order to remove sediment, which, if allowed to accumulate, will cut the bronze plug and cause trouble. IT MUST ALWAYS BE DRAINED AT NIGHT AND WHEN NOT IN USE IN COLD WEATHER, in order to prevent damage to parts by freezing.

Mixer on Work

After mixer has been taken to the work, we suggest putting one or more wheelbarrows full of stone in the drum and running the mixer for half an hour or more, thus scouring it thoroughly and removing all roughness. The revolutions of the drum should be tried out. The best results are obtained at the following revolutions per minute:

No. of Mixer	R.P.M.
104S	17
10S, 14S, 214E, 10E and 14E	16
107S, 21S and 21E	15
28S and 32E..............................	14

The pickup buckets, discharging directly upon the swing chute, also give the quickest discharge at the above stated revolutions. At a lower speed a certain amount falls off the buckets before reaching the proper height and at a greater speed some of the concrete is carried over the swinging chute.

It is advisable, upon starting a new mixer, to use one-half the normal crew the first day, in order to allow the operator to become acquainted with his machine and to make minor adjustments of the clutches.

After running the machine a couple of days, remove all slack from cables, as new cables stretch with use. This eliminates the jerks by which they are sometimes broken.

After running a week or so, all nuts should be drawn up, as the new bolts may stretch, leaving the nuts loose.

Operating MIXER TO INSURE MAXIMUM Yardage

The operator should throw in the hoisting clutch, then when the skip has reached the elevation where it is ready to charge drum, start the water flowing from the measuring tank. As automatic knockouts are on all Koehring Machines the skip will stop when reaching the proper height. After skip has been lowered the operator should close the three-way valve when the water stops flowing. He can tell this by watching the gauge glass on the water tank. When concrete has been mixed the required length of time, has been dumped, and three-fourths of the concrete is out of the drum, the skip should be started up again. By the time it has reached the charging position the drum is empty. Then reverse the tilting discharge chute so that it can aid in mixing the next batch. The operator will have time to run bucket out on the boom and dump it while the batch is being mixed and the skip is being loaded.

Keep Mixer Clean

When contractors finish concreting at noon and in the afternoon, it is advisable to put a few wheel-barrows of stone in the drum and scour it for a few minutes. By doing this a clean drum is assured at all times.

To help instill in the minds of the crew pride in their work and to secure a maximum of output, the machine should be kept as clean as possible, both on the inside and on the outside. The outside may be kept clean by brushing off the machine each night before shutting down and coating the drum and other parts with oil, which will prevent the cement from sticking. The oil drained from the nk case can be saved and used for this purpose.

In coating the drum and other parts of mixer with oil, care must be taken that no oil gets on the runways of either the drum or the boom.

The Proper Method of Putting Aggregate in Skip to PREVENT CONCRETE CLOGGING on Blades

Cement should never be put in skip first, as it retards the flow of material when skip is elevated, thereby making it necessary to hold it in that position for a longer time to clean itself. As the cement is last to leave the skip, some of it will stay in the sub-chute until pushed into the drum by the next batch, and some will pack on the inside of the drum head on the charging side, building up a ring around the drum opening. There should always be at least $\frac{1}{4}$ of the stone or gravel in skip before cement is dumped, as this will aid the flow of material and also prevent the cement hanging in the sub-chute.

It is very important that the water be admitted at the proper time; that is, when the material starts to roll into the drum, so the material and water will enter the drum at the same time to avoid clogging. This also increases the mixing efficiency.

Shutting Down Mixer for Winter or Prolonged Period, When Machine Will not Be in Use

Drain all water from the boiler and steam engine or gasoline engine, and also from piping and water tanks. Pour about one gallon of cylinder oil into the boiler and a quart of oil into the tanks. Fill again with water, then drain. It would be well to add a little whitelead to the cylinder oil,—just enough to give it a little body.

If mixer is equipped with steam power, remove hood from boiler and clean the tubes out thoroughly with a wire brush and swab carefully with oiled waste or rags. Cover the top of the boiler with canvas and tie it down securely, then coat inside of fire box with heavy oil, and paint or oil the outside of the boiler.

Whether gasoline or steam engine is used, take

off cylinder head and coat inside of cylinder with whitelead and oil.

All hard working machinery, especially concrete machinery, must wear; therefore examine the entire machine carefully and try out all working parts and replace with new parts those parts, if any, which are worn enough so as not to work properly.

By overhauling the mixer and replacing all worn parts, when mixer is shut down, no delay will be occasioned by having to make repairs when ready to re-start the mixer in the spring. This should not be neglected, for worn machinery decreases results and increases delays and consequently labor, and possibly will necessitate extra hours at night for the operator "when the mosquitoes are biting at their best."

After machine has been thoroughly inspected and overhauled, coat all bright parts on mixer with heavy machine oil or, preferably, cylinder oil containing a good heavy body of whitelead. Give balance of mixer a good coat of paint.

Grease all bearings carefully and turn machinery over a few times to insure that the insides of bearings are thoroughly coated. Take off cables and clean thoroughly with kerosene, then pass through a bath of warm lubricant. The following makes a good mixture:

One part freshly slacked lime.

Four parts fine or coal tar.

One-half part powdered graphite.

Heat up and thoroughly mix together.

Clean traction chain thoroughly with kerosene, then give it a good coat of heavy oil.

If possible, house the mixer or cover it with tarpaulin in order to protect it from the weather.

When starting up again after a prolonged shut down, it will, of course, be necessary to clean off all surplus grease and then follow the same procedure as used in starting of a new machine.

Ordering Parts

When ordering new parts for a mixer, the contractor should be sure to give the size and number of his machine. This he will find on the nameplate attached to the frame. The part and pattern number he will find in the repair part book which accompanies each mixer.

Ordering Clutch Parts

When ordering new parts for clutches, state specifically whether parts are wanted for—

Drum or main drive clutch.
Charging skip hoist clutch.
Traction clutch.
Boom clutch or auxiliary hoist clutch.

AND ALWAYS GIVE THE NUMBER OF YOUR MIXER.

(Illustration)

Part No. 65, master disc for charging skip hoist clutch on mixer.

Size ....No. ....
Get these numbers from nameplate on your mixer

Or

.... Mixer No........
code word of part   code word of HP clutch

This will eliminate mistakes and will enable us to give our customers prompt service.

INDEX

INDEX TO TABLES

INDEX TO TABLES—Continued

LIST OF ILLUSTRATIONS

CPSIA information can be obtained
at www.ICGtesting.com
Printed in the USA
BVHW04*1032310718
523166BV00008B/115/P

9 781356 828319